Swarm Intelligence for Resource Management in Internet of Things

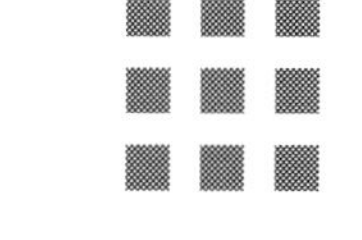

Swarm Intelligence for Resource Management in Internet of Things

Edited by

Aboul Ella Hassanien
Cairo University, Cairo, Egypt

Ashraf Darwish
Faculty of Science, Helwan University, Cairo, Egypt

Series editor
Fatos Xhafa

Academic Press is an imprint of Elsevier
125 London Wall, London EC2Y 5AS, United Kingdom
525 B Street, Suite 1650, San Diego, CA 92101, United States
50 Hampshire Street, 5th Floor, Cambridge, MA 02139, United States
The Boulevard, Langford Lane, Kidlington, Oxford OX5 1GB, United Kingdom

British Library Cataloguing-in-Publication Data
A catalogue record for this book is available from the British Library

Library of Congress Cataloging-in-Publication Data
A catalog record for this book is available from the Library of Congress

ISBN: 978-0-12-818287-1

For Information on all Academic Press publications
visit our website at https://www.elsevier.com/books-and-journals

Publisher: Mara Conner
Acquisitions Editor: Chris Katsaropoulos
Editorial Project Manager: Gabriela D. Capille
Production Project Manager: Joy Christel Neumarin Honest Thangiah
Cover Designer: Victoria Pearson

Typeset by MPS Limited, Chennai, India

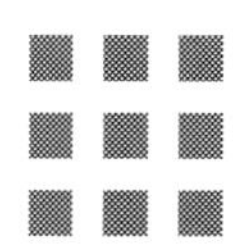

Contents

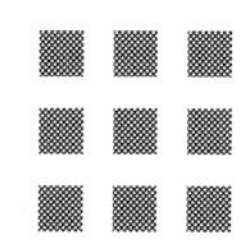

List of contributors

Amr Abohany Information Systems Department, Faculty of Computers and Information Systems, Kafrelsheikh University, Kafr El Sheikh, Egypt

Ahmed F. Ali Department of Computer Science, Faculty of Computers & Informatics, Suez Canal University, Ismailia, Egypt

Mazoon AlRubaiei Modern College of Business and Science, Muscat, Oman

Ashraf Darwish Faculty of Science, Helwan University, Cairo, Egypt

S. Dhamodharavadhani Department of Computer Science, Periyar University, Salem, India

Engy El-Shafeiy Computer Science Department, Faculty of Computers and Artificial Intelligence, Sadat City University, Sadat City, Egypt

Hothefa sh Jassim Modern College of Business and Science, Muscat, Oman

Mazen Juma Faculty of Engineering and Information Technology, British University in Dubai, Dubai, UAE

Avinash Kaur School of Computer Science and Engineering, Lovely Professional University, Phagwara, India

Fahad Layth Malallah Computer and Information Technology, Collage of Electronic and Electrical Engineering University, Baghdad, Iraq

Elezabeth Mathew BUiD, Dubai, United Arab Emirates

Anand Nayyar Duy Tan University, Da Nang, Viet Nam

Chhabi Rani Panigrahi Department of Computer Science, Rama Devi Women's University, Bhubaneswar, India

Bibudhendu Pati Department of Computer Science, Rama Devi Women's University, Bhubaneswar, India

Binod Kumar Pattanayak Department of Computer Science and Engineering, Siksha 'O' Anusandhan (Deemed to be University), Bhubaneswar, India

Aliaa R. Raslan Higher Institute for Management Information Systems, Suez, Egypt

Mamata Rath School of Management (IT), Birla Global University, Bhubaneswar, India

R. Rathipriya Department of Computer Science, Periyar University, Salem, India

Sohail Safdar College of Information Technology, Ahlia University, Manama, Bahrain

Khaled Shaalan Faculty of Engineering and Information Technology, British University in Dubai, Dubai, UAE

Baraa T. Sharef College of Information Technology, Ahlia University, Manama, Bahrain

Zeyad T. Sharef Faculty of Engineering, University of Auckland, Auckland, New Zealand

Parminder Singh School of Computer Science and Engineering, Lovely Professional University, Phagwara, India

S. Sivabalan Department of Computer Science, Periyar University, Salem, India

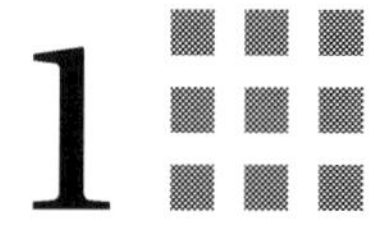

Swarm intelligence algorithms and their applications in Internet of Things

Aliaa R. Raslan[1], Ahmed F. Ali[2], Ashraf Darwish[3]

[1]HIGHER INSTITUTE FOR MANAGEMENT INFORMATION SYSTEMS, SUEZ, EGYPT [2]DEPARTMENT OF COMPUTER SCIENCE, FACULTY OF COMPUTERS & INFORMATICS, SUEZ CANAL UNIVERSITY, ISMAILIA, EGYPT [3]FACULTY OF SCIENCE, HELWAN UNIVERSITY, CAIRO, EGYPT

1.1 Introduction

The Internet of Things (IoT) is considered to be one of the most important technologies that will make a huge change to the world. IoT means that everything (both digital and physical objects) around us is connected via the internet and has the capability to sense, process, connect, and operate. This allows new classes of services and applications [1]. A lot of challenges existing in the IoT need to be solved to achieve such applications and services.

Swarm intelligence (SI) has been inspired by the collective conduct of local interaction between insects or animals like termites, ants, cats, fishes, or birds [2,3]. Many SI algorithms have been proposed and used in efficient manner in a large scope of problems and proved their success in resolving the world's problem. SI is an important example of algorithms that can transact complicated issues like system problems of the IoT. We can represent systems of IoT like a swarm of devices and combine it with SI algorithms to overcome a lot of challenges [4].

IoT devices work by batteries that are characterized by a limited and nonrenewable energy supply. These devices consume energy during sensing, processing, and transmitting data which leads to a reduction in the IoT network lifetime. Therefore, the reduction of energy consumption is considered to be the main challenge in the IoT and it is necessary to expand the lifetime for the IoT network. Clustering is an important technique that can minimize the consumption of energy for the IoT network. The selection of a cluster head (CH) in a wireless sensor network (WSN) is considered to be a serious method in order to conserve energy in the network. In this work, we review the implementations related to the CH process using metaheuristic algorithms for improving energy awareness in IoT and expanding the network lifetime.

Swarm Intelligence for Resource Management in Internet of Things. DOI: https://doi.org/10.1016/B978-0-12-818287-1.00003-6

The chapter includes the following sections: Sections 1.2 and 1.3 represent the meaning of SI algorithms and the IoT, respectively. Section 1.4 reviews the use of recently proposed SI algorithms to achieve the efficient CH selection in the IoT systems and thus save more energy and expand the network lifetime. Finally, the paper concludes in Section 1.5.

1.2 Swarm intelligence techniques

The origins of the term SI date back to the 1980s and Beni, who presented a category of cellular robot systems. This was used in different area of studies for optimization problems [5].

The meaning of SI in accordance with Collins Dictionary is "an artificial intelligence (AI) approach for solving the issues by algorithms, these algorithms depend on the collective conduct of insects that organized by themselves" [4]. Any SI system consists of a colony of simple agents (such as an individual ant in an ant colony), and each agent in the colony interacts with their neighbor and their environment to accomplish their goals. As a result, together they can accomplish larger goals and solve real-world problems. According to computer science domain, SI can be defined as a collection of algorithms and concepts, which present and formalize this intelligence behavior. From the point of view of AI or computer science, the SI definition can be as follows: "swarm intelligence is a set of intelligent systems inspired by the collective intelligence of a group. This collective intelligence is accomplished within the direct or indirect interactions between agents which are homogeneous in nature, yet collaborate with each other in their local environment without taking into account the global pattern."

1.2.1 Swarm intelligence and nature-inspired algorithms

Many SI and nature-inspired (NI) algorithms that depend on the collective conduct of natural swarms have been developed to solve many problems. These algorithms mimic the insects [6–11], birds [12,13], animals [14–19], simple organisms [20,21], fishes [22–25], and mammals, when they search for food or mates, by sharing their information to reach the goal. They have been classified into categories as shown in Fig. 1–1.

1.2.2 Key characteristics for swarm intelligence systems

Any SI-based system should include the following three properties: self-organization (SO), stigmergy, and division of labor:

- *SO*: considered as one of the most important properties of SI systems. SO is one global behavior or the property of SI systems that is accomplished by the interactions between its lower level agents; these interactions rely on a collection of rules. The key element of these rules is that there is no need for external governance to control the local behavior of agents.
- *Stigmergy*: the interaction rules should interact with any change in the environment and the agent should be able to adapt to this change autonomously and continue to achieve its function.

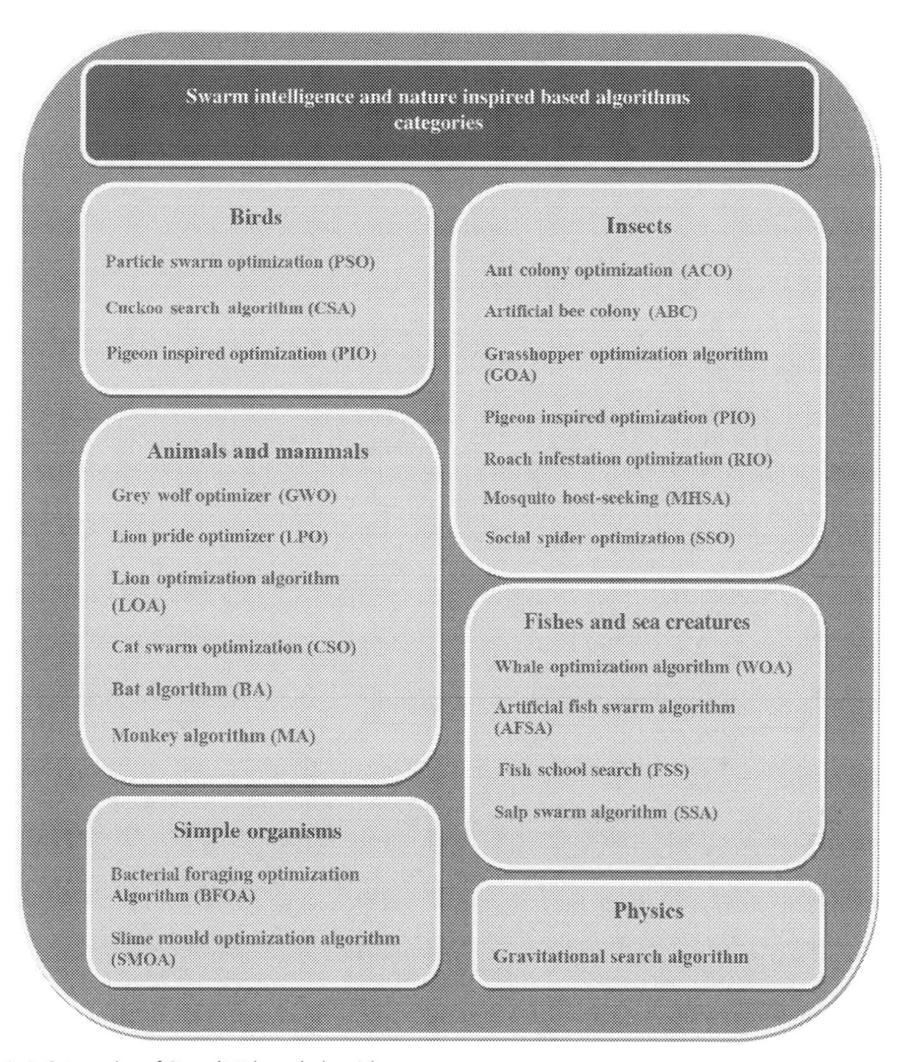

FIGURE 1–1 Categories of SI and NI-based algorithms.

- *Division of labor*: each agent in the swarm has a limited capability to achieve the goal for the entire swarm. For this reason, the natural system uses the division of labor with agents to execute a set of specific functions that lead to the overall success of the swarm.

1.3 Internet of Things

Over 15 years ago, a significant new title or term was christened called the "Internet of Things." This term originates at the Massachusetts Institute of Technology (MIT) [26,27] and after that, the term has been improved and expanded behind the range of radio frequency indentification (RFID) technologies. IoT expands internet connectivity behind conventional devices such as tablets, smartphones, personal computers, and laptops to a continuously growing network of everyday objects/things. This network uses distinctive technology for connecting and reacting with the environment and enables things to aggregate and interchange data via the internet. The IoT has the ability to connect many resources to each other to provide smart services in various applications [28]. A large number of organizations in different industries utilize IoT increasingly to work in an efficient manner, that is, to understand clients in a better way to deliver improved customer service, to take decisions, and to enhance business values. These achievements have proved the effectiveness and promising future of IoT in different application areas.

1.3.1 Internet of Things definitions

A huge number of definitions for the IoT have been proposed, these definitions include three types [29]. The first type includes the definitions that focus on things that are associated in the IoT. The second type includes definitions that are interested with the correlated concepts for the internet, like protocols and technology of networks. Challenges of the IoT, like searching, storing, and organizing huge amounts of information have been involved in the final type [26]. The IoT can be defined in different forms:

- IoT is defined as the system, which is associated with devices, people, animals, mechanical, and digital tools or another different thing. All of these components have unique identities and can convey data via the network with no need for humans [30–33].
- IoT is a network consisting of a large number of sensors, appliances, and buildings connected to the internet to generate, gather, and interchange enormous volumes of data [6,34].
- IoT means "the global infrastructure for the knowledge community that enabling developed services by interlink between physical and virtual things depending on current and developing, interoperable information and communication technologies" according to the International Telecommunication Union (ITU) [35].
- IoT is powerful connectedness among the digital and the real world [26].
- IoT is "a dynamic global network infrastructure having self-configuring abilities depend on conventional and interoperable communication protocols where physical and virtual things have virtual characters and physical properties also utilize smart interfaces, and can be combined easily into the information network" [36].
- IoT is "objects owning identities and virtual personalities working in smart areas by utilizing smart interfaces to connect and interact inside the environmental, social, and user contexts" [37].
- According to the semantic concept "IoT refers to a worldwide network of things that communicate with each other and can deal with it in a unique way depending on conventional communication protocols" [38].
- From the concept of machine-to-machine (M2M), "IoT defined as a network contains billions of intelligent devices which connect with individuals and several applications to gather and participate data also M2M presents the connection that allows IoT" [39].

- According to the European Union, "IoT means a network of things/objects, these things can connect with users, society and environment" [40].

1.3.2 Key characteristics for Internet of Things

There are a lot of characteristics for the IoT which vary from one domain to another, some of the general and key characteristics include [41]

- *Intelligence*: IoT consists of the combination of algorithms and computation, software and hardware that makes it smart. Intelligence in IoT improves its capabilities, which enable the things to respond in an intelligent way to a specific situation and help them to execute specific tasks.
- *Connectivity*: anything in the IoT, like sensors, devices, and actuators, can be connected with the global information and each other.
- *Dynamic changes*: the devices of the IoT can change their state dynamically, which means sleeping and wake up, connecting and disconnecting, and also the location, speed, and the number of devices can be changed.
- *Things*: includes everything that can be connected such as sensors and devices.
- *Enormous scale*: the number of appliances that connect with each other and need to be managed will be greater than the appliances connected to the internet at present.
- *Heterogeneity*: IoT devices depend on various networks and hardware platforms, these devices can react with another device or platform within diverse networks.

1.3.3 Internet of Things applications

There are many applications for IoT technologies, such as the smart industry, which improve the systems that involve intelligent production. Another application called smart home focuses on the systems that are responsible for security and thermostats. Smart energy includes the smart counters of water, gas, and electricity. One more application called smart transport is mobile ticketing and vehicle fleet tracking. A further application called smart health is the observation of patients. Smart city projects include intelligent lighting of streets and monitoring of parking space availability [6,42,43].

1.4 Proposed swarm intelligence models

In the following subsections, we highlight the most important IoT problems and the solutions of these problems using SI algorithms.

1.4.1 Proposed clustering model 1

1.4.1.1 Problem definition

Energy awareness is considered the major challenge for the IoT [44–47]. Clustering is considered a significant technique that is used for expanding the lifetime in a WSN. However, in some cases CH may deplete extra energy due to the additional load to receive and collect the packets of data from other nodes in their clusters and then send the collected packets to base station [48]. For this reason, the proper CH selection is an important factor in order to conserve energy, leading to the expansion of the lifespan of the network. A novel method consisting of an integration of artificial bee colony (ABC) and gravitational search algorithm (GSA) algorithms is proposed in Ref. [49] to elect the CH in an effective way in the IoT domain.

1.4.1.2 Network model

The proposed model depends on the following assumption:

- N number of IoT devices and sensor nodes are distributed across the $(M \times N)$ region in meters.
- Each sensor node is connected with a separate IoT device in order to monitor and send data to the IoT device as shown in Fig. 1–2.
- The selection of CH takes place between three clusters, these clusters include many IoT devices. Then CHs collect data from the rest of the devices and transmit the data to the base station I_B.
- A, B, C represent the three selected CHs.
- C_{In} represents the clusters of the network while H_{In} represents the CH.
- D_{mn} represents the distance among the mth devices and the nth device.
- D_{HIB} means the distance among the CH and the base station.

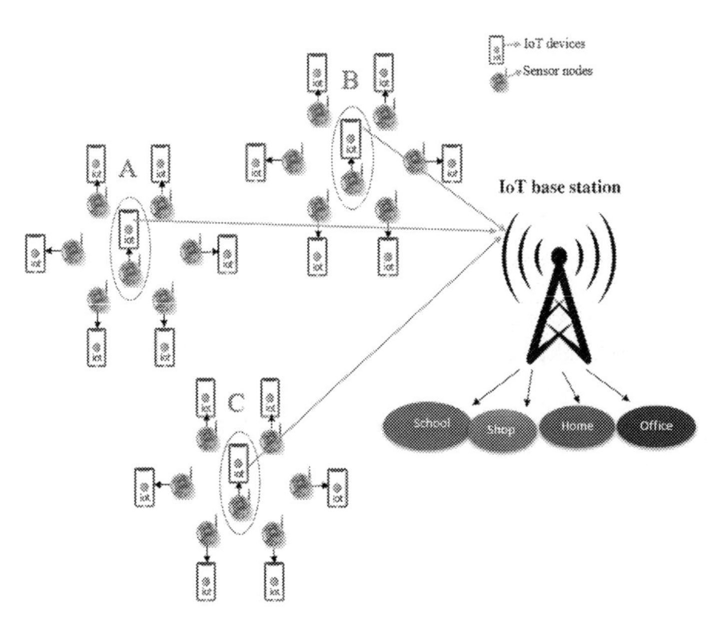

FIGURE 1–2 IoT network.

1.4.1.3 Selection of cluster head

The selection of the CH in the IoT network depends on some parameters, including load, temperature, energy, distance, and the delay for IoT devices. All of these parameters must decrease, except for energy that must increase. Therefore the fitness function is a maximization function according to Eqs. (1.1–1.3)

$$OF_1 = \frac{O_f^{\text{energy}}}{O_f^{\text{load}}} + \frac{O_f^{\text{energy}}}{O_f^{\text{temperature}}} \tag{1.1}$$

$$OF_2 = \frac{\beta}{O_f^{\text{distance}}} + (1-\beta)OF_1 \tag{1.2}$$

$$OF_3 = \gamma OF_2 + \frac{(1-\gamma)}{O_f^{\text{delay}}} \tag{1.3}$$

where β, γ are constants with values of 0.9 and 0.3, respectively.

1.4.1.4 The proposed artificial bee colony and gravitational search algorithm

The proposed algorithm consists of the GSA algorithm combined with the ABC algorithm. This algorithm is used for the selection of CH in IoT network; it uses the upgrade step of the employed bee stage of ABC algorithm, so by the ABC algorithm the velocity will be upgraded as shown in Eq. (1.4)

$$V_i^d(t+1) = V_i^d(t) + Q_i(V_i^d(t) - V_j^d(t)) + A_i^d \tag{1.4}$$

where V_i^d is the current velocity of agent I. Q_i is a random number in the range $[-1, 1]$. V_j^d is the neighbor velocity (Algorithm 1.1).

Algorithm 1.1 Proposed GSA algorithm

1: Generate the initial population of the agents $m = 1, 2, \ldots, N$.
2: **repeat**
3: Compute the values of M, $G(t)$, $B(t)$, and $W(t)$.
4: Generate the initial position and velocity $X_i^d(t)$, $V_i^d(t)$ of all agents, respectively.
5: Evaluate all agents by calculating the fitness function of them.
6: Assign the Kbest agents.
7: Calculate the force and acceleration for each agent.
8: Update the position and the velocity for all the agents.
9: **until** Termination criteria satisfied.

1.4.2 Proposed clustering model 2

1.4.2.1 Problem definition

One of the most significant components in the IoT is called the big data sensing system (BDSS). The reduction of energy consumption is an important challenge for the IoT. A lot of energy consumption algorithms have been improved and used for BDSS. One of them is the low energy adaptive clustering hierarchy (LEACH) protocol [50] that is used for WSN. Although this protocol expands the lifetime of the network, it needs much energy for the data transmission. As a result, the nodes may die earlier, thus leading to the ending of the network lifetime. A new method that consists of a combination of bat algorithm and centroid strategy is proposed in Ref. [51]. This method includes several designs for three centroid strategies. One of them consists of a weighted harmonic centroid strategy combined with bat algorithm and is called (WHCBA). This algorithm has proved that it is the best of the other centroid strategies that were combined with the bat algorithm and then applied to LEACH protocol to develop the LEACH performance. This algorithm is called LEACH-WHCBA.

1.4.2.2 Combination of bat algorithm and centroid strategy

The global and local search patterns in the bat algorithm prevent entry to some neighboing areas and limit their searching abilities. A new method called the centroid strategy is designed to solve this problem. This strategy takes into consideration all the consequences resulting from the best position $P(t)$ and neighbor bats. For clarification: there are three bats selected. The first of them is close to the best position and the other two bats are far away from the best position. The position gained from centroid strategy is used to exchange the bad position of one bat. Also, we can observe that the local searching area can expand by using the centroid strategy. The main steps of the modified algorithm are written in Algorithm 1.2.

The modified bat algorithm involves different designs for three centroid strategies as follows:

- Arithmetic centroid strategy, which is calculated as follows:

$$X_j(t+1) = \frac{\sum_{i=1}^{m} X_i(t)}{m} \tag{1.5}$$

  where $X_i(t)$, $i = 1, 2, \ldots, m$ represent the bats that have performed better than $X_j(t)$,
- Geometric centroid strategy

$$X_j(t+1) = \sqrt[m]{\prod_{i=1}^{m} X_i(t)} \tag{1.6}$$

- Harmonic centroid strategy

$$X_j(t+1) = \frac{m}{\sum_{i=1}^{m} \frac{1}{X_i(t)}} \tag{1.7}$$

Algorithm 1.2 Modified bat algorithm with centroid strategy

1: Initialize the position, velocity, and parameters for each solution (bat) in the population.
2: **repeat**
3: Generate the frequencies for each solution randomly.
4: Calculate the velocity for each solution as shown in Eq. (1.12)
5: **if** $rand < r_i$ **then**
6: Calculate the temp position for each solution.
7. **else**
8: **if** current generation $< \lambda\%$ of total generation **then**
9: Calculate the temp position for the corresponding solution with one centroid strategy.
10: **else**
11: Calculate the temp position for corresponding solution.
12. **end if**
13. **end if**
14: Calculate the position and velocity for solution s as shown in Eqs. (1.13) and (1.14).
15: Evaluate the quality (fitness) for each solution.
16: Update the position for the current solution.
17: Update the loudness and emission rate.
18: Evaluate the solutions and save the best position.
19: **until** Termination criteria satisfied.
20: Produce the best position;

The weighted counterparts for Eqs. (1.8–1.10)

- The weighted arithmetic centroid strategy

$$X_j(t+1) = \sum_{i=1}^{m} w_i X_i(t) \tag{1.8}$$

- The weighted geometric centroid strategy

$$X_j(t+1) = \sqrt[m]{\prod_{i=1}^{m} X_i(t)^{w_i}} \tag{1.9}$$

- The weighted harmonic centroid strategy

$$X_j(t+1) = \frac{m}{\sum_{i=1}^{m} \frac{w_i}{X_i(t)}} \tag{1.10}$$

where w_i is the weight that depends on the bat performance

$$w_i = \frac{e^{\text{score}_i(t)}}{\sum_{j=1}^{n} e^{\text{score}_i(t)}} \tag{1.11}$$

$$\text{score}_j(t) = \begin{cases} \dfrac{f_{\text{worst}} - f_j}{f_{\text{worst}} - f_{\text{best}}} & \text{if} \quad (f_{\text{worst}} > f_{\text{best}}) \\ & \text{otherwise} \end{cases}$$

1.4.2.3 Velocity inertia-free strategy

The global search ability in the standard bat algorithm has been represented by using the equation of velocity update which includes the effect of $P(t)$ and the inertia part $V_{ik}(t)$. Because the velocity inertia part limits the bat movement where a bat with large inertia hardly moves, so the velocity update is calculated in the following equation:

$$V_{ik}(t+1) = (X_{ik}(t) - P_k(t)) \cdot Fr_i(t) \tag{1.12}$$

where the inertia part has been removed. For this reason, the bat flies in a more flexible way. The best position $P(t)$ found in the last generations, will become the position at present for one bat s in accordance with the position upgrade rule, so bat s upgrades its position and its velocity will be as following:

$$V_{sk}(t+1) = X_{min} + (X_{max} - X_{min}) \cdot rand \tag{1.13}$$

$$V_{sk}(t+1) = X_{sk}(t+1) - X_{sk}(t) \tag{1.14}$$

where $[X_{min}, X_{max}]$ represents the search domain for each coordinate.

1.4.2.4 Improved low energy adaptive clustering hierarchy with weighted harmonic centroid bat algorithm

The weighted harmonic centroid bat algorithm (WHCBA) has been applied to the LEACH protocol to develop the performance of the LEACH protocol. The algorithm is called LEACH-WHCBA. The selection of the CH in this algorithm consist of two steps called "temporary" and "formal" CH selection. The first step includes the conventional LEACH protocol. In the second step, the WHCBA has been used to improve the temporary selection that happened in the first step. The best position is obtained by calculating the distance between each node in the same cluster and the base station and then calculating the variance of these distances to find the minimum which represents the best position. So, the objective function for this algorithm is calculated in the following equation:

$$f = \alpha_1 \cdot std\left(\sum_{i=1}^{k} D_{iN}\right) + \alpha_2 \cdot D_S \tag{1.15}$$

where k represents the nodes number in cluster c_j, D_S means the distance between the individual bat and the base station, D_{iN} is the distance between the individual bat and node i inside the cluster, and α_1 and α_2: are user-defined constants and $\alpha_1 + \alpha_2 = 1$.

The residual energy average for nodes in the present cluster has been calculated as:

$$Ave_E = \frac{\sum_{i=1}^{k} s(i) \cdot E}{k} \tag{1.16}$$

where $s(i) \cdot E$ is the residual energy for node i.

This fitness function aims to minimize the distance between each node inside the cluster and its CH and the base station. The formal node means the node nearest to the best position and is chosen from other nodes that have energies greater than the cluster's average energy. As a result, the developed LEACH algorithm decreases the consumption of energy for each node and has more remaining energy.

1.4.3 Proposed clustering model 3

1.4.3.1 Problem definition

The devices in the internet of things for mobile ad-hoc networks (MANETs) can be divided into clusters. These clusters are connected to each other and have the ability to send data to the destination. Also, the topology of the network can change dynamically. One of the challenges of clustering in MANETs for IoT is reducing the size of the routing table and reduces the topology maintenance overhead. A new method is proposed [52] by combining the characteristics of genetic algorithms (GAs) with honey bees to assist the population to handle the dynamic changes and solve these problems.

1.4.3.2 Network model

The proposed model depends on the following assumptions:

- The MANET for IoT is an undirected and connected topology graph $G(V, E)$ where V represents the nodes group while E represents the communication links set among the nodes in the same communication range.
- The network consist of n number of nodes (communication devices) that connect with each other.
- K represents the number of CHs that need to be selected from the network nodes where the volume of the CHs group must be as low as possible and the member nodes number should be nearly the same.
- The distance should be nearly the same for each CH to other CH.
- To conserve more energy, some of the nodes move to sleep state and the others wake up randomly or cyclically.
- Topology is not fixed, it changes dynamically because the nodes shift from one position to another, and we need to obtain the cluster headset in a fast way when there is a change in the topology.

1.4.3.3 Selection of cluster head

Choice of CH in the IoT for MANETs rely on some parameters such as energy (E_{node}), degree (D_{node}), mobility (M_{node}), and quality (Q_{node}) of the node (communication device).

Energy means the node (communication device) that has more energy than the other nodes has a great opportunity to be a CH. The degree means the communication device that has the highest number of neighbors is the optimum to be a CH. Mobility means the communication devices that have proportional mobility are the best candidates to be a CH. Mobility depends on the communication device's speed and direction. Quality includes a good neighbor, which means a neighbor with a one-hop distance; a satisfactory neighbor means a neighbor with a two-hop distance; otherwise they are not neighbors. For each communication device *i*, its fitness (weight) is computed as follows:

$$X_i = E_{\text{node}} + Q_{\text{node}} + D_{\text{node}} + M_{\text{node}} \tag{1.17}$$

The trouble of collecting *n* of communication devices of the ad-hoc network inside *k* clusters is considered in Eq. (1.18) according to [53].

$$\text{Minimize} \quad F(W, C) = \sum_{i=1}^{n} \sum_{j=1}^{k} w_{ij}(x_i - c_j)^2, \tag{1.18}$$

$$\text{Subject to} \quad \sum_{j=1}^{k} w_{ij} = 1 \quad i = 1, 2, \ldots, n,$$

$$w_{ij} = 0 \text{ or } 1 \quad i = 1, 2, \ldots, n \quad j = 1, 2, \ldots, k,$$

where x_i, $i = 1, \ldots, n$ represents weight for device *i*, c_j, $j = 1, \ldots, k$ represents the node's average fitness and is computed as follows:

$$c_j = \frac{1}{N_j} \sum_{j=1}^{k} w_{ij} x_i \tag{1.19}$$

where N_j nodes number in the *j*th cluster. The relationship weight for a communication device *i* which exists in cluster *j* is obtained by w_{ij}. If the device is a member of cluster *j* then $w_{ij} = 1$, otherwise $w_{ij=0}$.

Hence, the new CH selection depends on its probability related to the nectar amount so the next equation shows the probability that onlooker bees will visit the CH:

$$p_i = \frac{Q(Nec_i)}{\sum_{j=1}^{fs} Q(Nec_j)} \tag{1.20}$$

where $Q(Nec_i)$ is the amount of nectar at communication device *i*, and *fs* is the CH's total number.

The following equation represents the cost function that needs to be minimized:

$$cf_i = \frac{1}{k} \sum_{j=1}^{k} d(x, CH_j) \tag{1.21}$$

where d represents distance among CH j and its neighbors, whether neighbors with one or two hops, CH_j represents the CH j in the set of CHs, and x is a random node which refers to the CH's neighbors.

1.4.3.4 Proposed clustering algorithm for genetic bee tabu

The genetic bee tabu (GBTC) algorithm works as follows: the weights of nodes are computed in accordance with Eq. (1.17), after that a group of nodes "scout bees" are selected that have a greater weight to be CHs. According to Eq. (1.18) the objective of the CHs group is calculated. Then the steps in the algorithm from step 3 to step 8 are execute; in step 5 the information about the CHs that have been selected is broadcast to the rest of nodes in the network. There are two stages in the proposed method: the cluster setup stage and the maintenance stage. The steps of the GBTC algorithm have been written in Algorithm 1.3.

- *Cluster setup phase* is responsible for finding the appropriate group of CHs $c_j(j = 1,\ldots, k)$ which decrease the objective function in Eq. (1.18). In the beginning the population of c scout bees has been used, where every scout bee refers to a potential CH. At first, the CHs group has been selected depending on its weights. According to the fitness function, each scout bee's fitness has been calculated. Then according to the best solution, the CHs are selected. To know the node number that the CH will serve, the relation between every

Algorithm 1.3 Genetic bee tabu clustering (GBTC)

1: Set the values of CHs, nodes (n), neighbors ns, mutation probability (pm), set of CHs c.
2: Set the tabu list s.
3: Generate a population with scout bees (c_j, $j = 1, 2, 3,\ldots, n$).
4: **if** fitness using Eq. (1.18) satisfied **then**
5: Get the optimal solution.
6: **else**
7: **repeat**
8: Calculate m nodes for neighborhood where $m = n - c$.
9: Compute the weights of m nodes using Eq. (1.17).
10: Select the best solutions for the new population.
11: **repeat**
12: Select two random solutions from $(n - m)$ the remaining solutions.
13: Apply the crossover operation on the selected two solutions.
14: Use pm on the offsprings.
15: Check offsprings in tabu list.
16: **until** number of offspring's $= n - m$.
17: Produce the new solutions.
18: **until** Termination criteria satisfied.
19: **end if**

node and every CH is specified. Bees that have greater weights have been chosen for the neighborhood search, and after that the neighborhood search is executed according to Eq. (1.22),

$$c_{j+1} = c_j + \lceil rand * n_s \rceil \tag{1.22}$$

where c_j represents the existing CH, c_{j+1} refers to other locations in c_j neighborhood, and n_s represents neighborhood volume. The remaining bees are assigned by using the GA.

The bee colony is produced in two forms: part of the colony has been produced by estimation of the fitness for the scout bees, while the rest of the colony has been produced by utilizing the GA in addition to the tabu search. Global and local searches in GBTC occur at the same time. The use of crossover and mutation operators makes the algorithm more efficient, as the crossover is characterized by its ability to replace solutions on the fly to produce better ones and the mutation achieves the diversity in population.

- *The maintenance phase* is where the MANET's topology for IoT is not fixed and changes dynamically. The maintenance phase is responsible for solving the changes that occur in the topology by reclustering the network when the intracluster communication links are lost. When a topology change happens the maintenance process begins locally on one node or group of nodes while the other nodes execute their ordinary operation. The most favorable CHs are produced again, which means "reclustering" without cutting the normal operations. One of the benefits of this algorithm is that every node can rapidly converge to the novel CH prior to the most favorable CH being obtained in the following iteration.

1.4.4 Proposed clustering model 4

1.4.4.1 Problem definition

In WSN-IoT network, the achievement of the proper CH selection is necessary to save more energy and prolong the network lifetime. Therefore, a novel method called "self-adaptive whale optimization algorithm" (SAWOA) is proposed in Ref. [49]. This algorithm takes into consideration the parameters of WSN such as distance, delay, and nodes energy and the IoT devices' parameters which include load and temperature. The proposed algorithm performance proves that it is more successful than particle swarm optimization, GA, adaptive GSA, ABC, GSA, and traditional whale optimization algorithm (WOA).

1.4.4.2 Network model

The proposed model of WSN-IoT network depends on the following assumption:

- N_c number of clusters
- C_i represent a cluster i, where $i = 1, 2, \ldots, N_c$
- S_{ij} refers to the node inside any cluster, where $j = 1, 2, \ldots, M$ and $i = 1, 2, \ldots, N$
- CH_i represents a CH, CHs are the only nodes in the network that can connect direct to the base station.

1.4.4.3 Selection of cluster head

The selection of a CH in WSN-IoT network depends on the WSNs parameters which include energy, delay, and distance between nodes and the parameters of the IoT network which include load and temperature, as shown in Fig. 1–3. To obtain an effective network the energy of the nodes must increase and the delay, temperature, distance, and load must decrease, and the energy should increase. Therefore the fitness function is calculated as shown in Eqs. (1.1), (1.2), and (1.3).

1.4.4.4 Proposed self-adaptive whale algorithm

In the proposed "self-adaptive whale optimization algorithm" (SAWOA) the position is updated using Eqs. (1.23) and (1.24). In Eq. (1.23), t is multiplied to a random selected k. t represents the fitness change as in Eq. (1.24). $f(t)$ refers to the present round and $f(t-1)$ refers to the prior round (Algorithm 1.4).

$$k = rand * t \tag{1.23}$$

$$t = \frac{f(t-1) - f(t)}{f(t-1)} \tag{1.24}$$

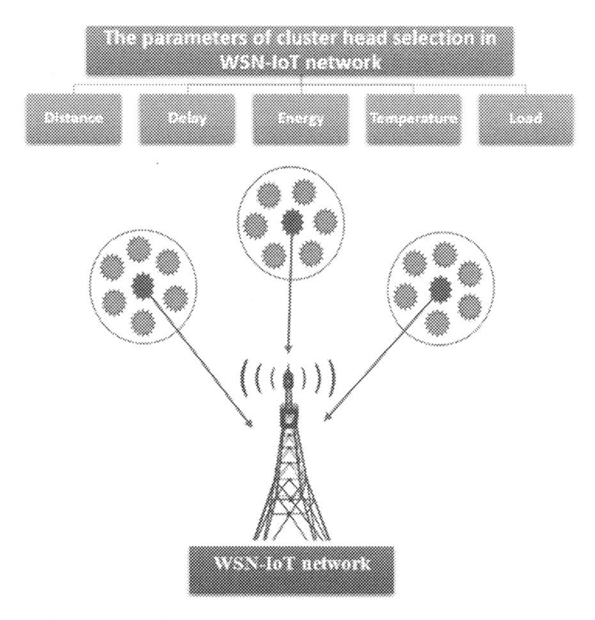

FIGURE 1–3 Clustering in WSN-IoT network.

Algorithm 1.4 SAWOA based cluster head selection

1: Generate the population of the agents $m = 1, 2,\ldots, N$.
2: Initialize the population of whales W_l, $l = 1, 2,\ldots, n$.
3: Set $t = 0$
4: Compute the fitness of each search agent.
5: Assign $W*$ as the best search agent.
6: **repeat**
7: **for** each search agent **do**
8: Update a, H, N and x.
9: Determine k using Eq. (1.23).
10: **if** $(k < 0.5)$ **then**
11: **if** $(|H| < 1)$ **then**
12. Update the position of current search agent.
13. **else**
14: **if** $(|H| \geq 1)$ **then**
15: Choose a random search agent W_{rand}.
16: Update the position of agents.
17: **else**
18: **if** $(|H| \geq 0.5)$ **then**
19: Update the position of agents.
20: **end if**
21: **end if**
22: **end if**
23: **end if**
24: **end if**
25: Test the presence of search agent moves apart from the search space and rectify it.
26. Compute the fitness of each search agent.
27: Update $W*$ for the condition of having better solution.
28: $t = t + 1$
29: **until** $t > t_{max}$.
30: Produce the best solution $W*$

1.5 Conclusion

Nowadays, most electronic devices are connected to the internet and they have sensors, processors, and actuators. These devices can form a network and share their information by collecting and sending data from their location to another location in the absence of humans. However this process consumes energy and decreases the network's lifetime. In this paper, we have given an overview of the recent SI algorithms which have been applied to solve this problem. The selection of a CH is one of the most used solutions for this problem. It can be considered as a global optimization problem and the SI algorithms were used to search for the optimal solution (the best CH).

References

[1] G. Fortino, P. Trunfio, Internet of Things Based on Smart Objects: Technology, Middleware and Applications, Springer Science & Business Media, 2014.

[2] S. Garnier, J. Gautrais, G. Theraulaz, The biological principles of swarm intelligence, Swarm Intell. 1 (1) (2007) 3–31.

[3] R.S. Parpinelli, H.S. Lopes, New inspirations in swarm intelligence: a survey, Int. J. Bio-Inspired Comput. 3 (1) (2011) 1–16.

[4] O. Zedadra, A. Guerrieri, N. Jouandeau, G. Spezzano, H. Seridi, G. Fortino, Swarm intelligence-based algorithms within IoT-based systems: a review, J. Parallel Distrib. Comput. 122 (2018) 173–187.

[5] B. Christian, M. Daniel, Swarm intelligence introduction and application, Nat. Comput. Ser. (2008).

[6] M. Dorigo, M. Birattari, T. Sttzle, Ant colony optimization artificial ants as a computational intelligence technique, IEEE Comput. Intell. Mag. 1 (2006) 28–39. Available from: https://doi.org/10.1109/MCI.2006.329691.

[7] H. Duan, P. Qiao, Pigeon-inspired optimization: a new swarm intelligence optimizer for air robot path planning, Int. J. Intell. Comput. Cybernet. 7 (1) (2014) 24–37.

[8] X. Feng, J.W. Zhang, H.Q. Yu, Mosquito host-seeking algorithm for tsp problem, Chin. J. Comput. 37 (8) (2014) 1794–1808.

[9] T.C. Havens, C.J. Spain, N.G. Salmon, J.M. Keller, Roach infestation optimization, 2008 IEEE Swarm Intelligence Symposium, IEEE, 2008, pp. 1–7.

[10] D. Karaboga, An idea based on honey bee swarm for numerical optimization (vol. 200), Technical report-tr06, Erciyes University, Engineering Faculty, Computer Engineering Department, 2005.

[11] S. Saremi, S. Mirjalili, A. Lewis, Grasshopper optimisation algorithm: theory and application, Adv. Eng. Softw. 105 (2017) 30–47.

[12] J. Kennedy, Handbook of nature-inspired and innovative computing, Swarm Intell. (2006) 187–219.

[13] X.S. Yang, S. Deb, Engineering optimisation by cuckoo search, Int. J. Math. Model. Numer. Optim. 1 (4) (2010) 330. http://dx.doi.org/10.1504/ijmmno.2010.035430.

[14] S.C. Chu, P.W. Tsai, Computational intelligence based on the behavior of cats, Int. J. Innov. Comput. Inform. Contr. 3 (1) (2007) 163–173.

[15] S. Mirjalili, S.M. Mirjalili, A. Lewis, Grey wolf optimizer, Adv. Eng. Softw. 69 (2014) 46–61.

[16] B. Wang, X. Jin, B. Cheng, Lion pride optimizer: an optimization algorithm inspired by lion pride behavior, Sci. China Inform. Sci. 55 (10) (2012) 2369–2389.

[17] Y.S. Yang, A new metaheuristic bat-inspired algorithm, Nature Inspired Cooperative Strategies for Optimization (NICSO 2010), Springer, Berlin, Heidelberg, 2010, pp. 65–74.

[18] M. Yazdani, F. Jolai, Lion optimization algorithm (LOA): a nature-inspired metaheuristic algorithm, J. Comput. Des. Eng. 3 (1) (2016) 24–36.

[19] R.Q. Zhao, W.S. Tang, Monkey algorithm for global numerical optimization, J. Uncertain Syst. 2 (3) (2008) 165–176.

[20] D.R. Monismith, B.E. Mayfield, Slime mold as a model for numerical optimization, in: 2008 IEEE Swarm Intelligence Symposium, 2008, pp. 1–8.

[21] K.M. Passino, Biomimicry of bacterial foraging for distributed optimization and control, IEEE Control Syst. Mag. 22 (3) (2002) 52–67.

[22] C.J. Bastos Filho, F.B. de Lima Neto, A.J. Lins, A.I. Nascimento, M.P. Lima, A novel search algorithm based on fish school behavior, 2008 IEEE International Conference on Systems, Man and Cybernetics, IEEE, 2008, pp. 2646–2651.

[23] X.L. Li, J.X. Qian, Studies on artificial fish swarm optimization algorithm based on decomposition and coordination techniques, J. Circuit. Syst. 1 (2003) 1–6.

[24] S. Mirjalili, A.H. Gandomi, S.Z. Mirjalili, S. Saremi, H. Faris, A.M. Mirjalili, Salp swarm algorithm: a bio-inspired optimizer for engineering design problems, Adv. Eng. Softw. 114 (2017) 163–191.

[25] S. Mirjalili, A. Lewis, The whale optimization algorithm, Adv. Eng. Softw. 95 (2016) 51–67.

[26] L. Atzori, A. Iera, G. Morabito, The internet of things: a survey, Comput. Netw. 54 (15) (2010) 2787–2805.

[27] M. Friedemann, C. Floerkemeier, From the internet of computers to the Internet of Things, Informatik-Spektrum 33 (2) (2010) 107–121.

[28] A. Darwish, A.E. Hassanien, Cyber physical systems design, methodology, and integration: the current status and future outlook, J. Ambient. Intell. Humanized Comput. 9 (5) (2018) 1541–1556.

[29] F. Wortmann, K. Flchter, Internet of things, Bus. Inf. Syst. Eng. 57 (3) (2015) 221–224.

[30] E. Kougianos, S.P. Mohanty, G. Coelho, U. Albalawi, P. Sundaravadivel, Design of a high-performance system for secure image communication in the Internet of Things, IEEE Access. 4 (2016) 1222–1242.

[31] Y. Liu, W. Han, Y. Zhang, L. Li, J. Wang, L. Zheng, An Internet-of-Things solution for food safety and quality control: a pilot project in China, J. Ind. Inf. Integr. 3 (2016) 1–7.

[32] G. Misra, V. Kumar, A. Agarwal, K. Agarwal, Internet of things (iot)-a technological analysis and survey on vision, concepts, challenges, innovation directions, technologies, and applications (an upcoming or future generation computer communication system technology), Am. J. Electr. Electron. Eng. 4 (1) (2016) 23–32.

[33] H. Park, H. Kim, H. Joo, J. Song, Recent advancements in the Internet-of-Things related standards: a oneM2M perspective, ICT Express 2 (3) (2016) 126–129.

[34] H. Ning, H. Liu, Cyber-physical-social-thinking space based science and technology framework for the Internet of Things, Sci. China Inf. Sci. 58 (3) (2015) 1–19.

[35] ITU, 2012. Internet of Things Global Standards Initiative.

[36] R.V. Kranenburg, The Internet of Things: A Critique of Ambient Technology and the All-Seeing Network of RFID, Institute of Network Cultures, 2008.

[37] D. INFSO, Networked Enterprise & RFID INFSO G. 2 Micro & Nanosystems, in co-operation with the Working Group RFID of the ETP EPOSS, Internet of Things in 2020, Roadmap for the Future [R]. Information Society and Media, Tech. Rep., 2008.

[38] P.R. Ray, A survey on Internet of Things architectures, J. King Saud. Univ.-Comput. Inf. Sci. 30 (3) (2018) 291–319.

[39] <https://internetofthingsagenda.techtarget.com/definition/Internet-of-Things-IoT>.

[40] A. Bassi, G. Horn, Internet of things in 2020, in: Proceedings of the Joint Euro-1049 Pean Commission/EPoSS Expert Workshop on RFID/Internet-of-Things, 2008.

[41] O. Vermesan, P. Friess, Internet of Things-from Research and Innovation to Market Deployment, vol. 29, River Publishers, Aalborg, 2014.

[42] E. Fleisch, What is the internet of things? An economic perspective, Econ. Manage. Finan. Mark. 5 (2) (2010) 125–157.

[43] O. Vermesan, P. Friess, P. Guillemin, H. Sundmaeker, M. Eisenhauer, K. Moessner, et al., Internet of Things Strategic Research and Innovation Agenda, vol. 7, River Publishers Series Communication, 2013.

[44] A. Karkouch, H. Mousannif, H. Al Moatassime, T. Noel, Data quality in internet of things: a state-of-the-art survey, J. Netw. Comput. Appl. 73 (2016) 57–81.

[45] S. Luo, B. Ren, The monitoring and managing application of cloud computing based on Internet of Things, Comput. Methods Prog. Biomed. 130 (2016) 154–161.

[46] A. Sivieri, L. Mottola, G. Cugola, Building internet of things software with eliot, Comput. Commun. 89 (2016) 141–153.

[47] T. Zhu, S. Dhelim, Z. Zhou, S. Yang, H. Ning, An architecture for aggregating information from distributed data nodes for industrial internet of things, Comput. Electr. Eng. 58 (2017) 337–349.

[48] P.S. Rao, P.K. Jana, H. Banka, A particle swarm optimization based energy efficient cluster head selection algorithm for wireless sensor networks, Wirel. Netw. 23 (7) (2017) 2005–2020.

[49] M.P.K. Reddy, M.R. Babu, Energy efficient cluster head selection for internet of things, N. Rev. Inf. Netw. 22 (1) (2017) 54–70.

[50] W.R. Heinzelman, A. Chandrakasan, H. Balakrishnan, Energy-efficient communication protocol for wireless microsensor networks, in: Proceedings of the 33rd Annual Hawaii International Conference on System Sciences, 2000, 10 pp.

[51] Z. Cui, Y. Cao, X. Cai, J. Cai, J. Chen, Optimal LEACH protocol with modified bat algorithm for big data sensing systems in Internet of Things, J. Parallel Distrib. Comput. 132 (2018) 217–229.

[52] M. Ahmad, A. Hameed, F. Ullah, I. Wahid, S.U. Rehman, H.A. Khattak, A bio-inspired clustering in mobile adhoc networks for internet of things based on honey bee and genetic algorithm, J. Ambient. Intell. Humanized Comput. (2018) 1–15.

[53] M. Ahmad, A.A. Ikram, R. Lela, I. Wahid, R. Ulla, Honey bee algorithm-based efficient cluster formation and optimization scheme in mobile ad hoc networks, Int. J. Distrib. Sens. Netw. 13 (6) (2017). P. 1550147717716815.

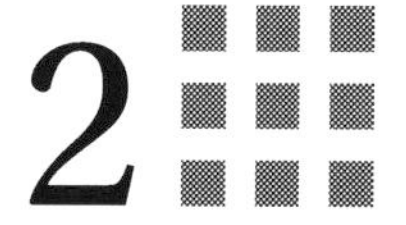

Swarm intelligence as a solution for technological problems associated with Internet of Things

Mamata Rath[1], Ashraf Darwish[2], Bibudhendu Pati[3], Binod Kumar Pattanayak[4], Chhabi Rani Panigrahi[3]

[1]SCHOOL OF MANAGEMENT (IT), BIRLA GLOBAL UNIVERSITY, BHUBANESWAR, INDIA [2]FACULTY OF SCIENCE, HELWAN UNIVERSITY, CAIRO, EGYPT [3]DEPARTMENT OF COMPUTER SCIENCE, RAMA DEVI WOMEN'S UNIVERSITY, BHUBANESWAR, INDIA [4]DEPARTMENT OF COMPUTER SCIENCE AND ENGINEERING, SIKSHA 'O' ANUSANDHAN (DEEMED TO BE UNIVERSITY), BHUBANESWAR, INDIA

2.1 Introduction

As per the technical aspect, swarm intelligence (SI) is an artificial intelligence (AI) technique based on collective behavior in decentralized, self-organized systems and it is generally made up of agents who interact with each other and the environment. Many people believe that it can progress technology, as it can tackle diversified issues from industrial applications to scientific research due to the dynamic algorithms offered by these advanced intelligence-based domains. This dynamic innovation of AI still experiences huge challenges, however, challenges can be tackled when other controlling scientific domains are integrated with it.

Most of the SI-based calculations have been created to address stationary advancement issues and, consequently, they can meet the ideal arrangements proficiently. Be that as it may, some certifiable issues have a dynamic domain that changes after some time. For such powerful advancement issues, it is hard for a regular SI calculation to follow the changing ideal once the calculation has combined on an answer. Over the most recent two decades, there has been a developing enthusiasm of tending to utilize SI computation because of its adjustment abilities.

In SI, there are no centralized control structures and it is based on group behavior found in nature. First introduced by Beni and Wang in 1989 with their study of cellular robotic systems, the concept of SI was expanded by Bonabeau, Dorigo, and Theraulaz in 1999. Swarm Robotics refers to the application of SI principles to collective robotics. It includes a group of simple robots that can only communicate locally and operate in a biologically inspired manner. This is a currently developing area of research.

Swarm Intelligence for Resource Management in Internet of Things. DOI: https://doi.org/10.1016/B978-0-12-818287-1.00005-X

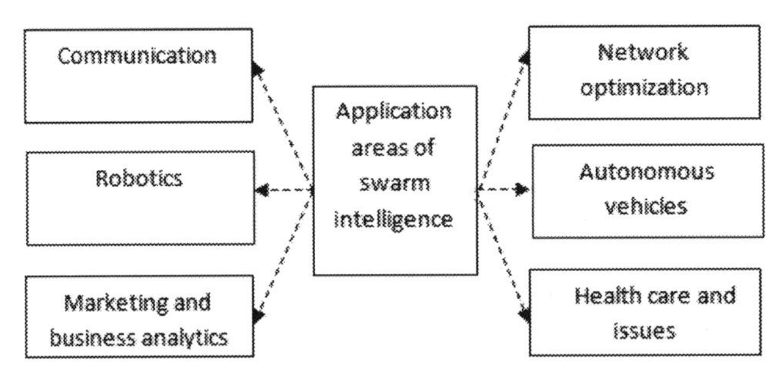

FIGURE 2–1 Different application areas of swarm intelligence.

With the rapid strengthening and extension of the Internet of Things (IoT) innovations in industrial applications, there was advancement of some related research concerns to understand the issues of their outcomes. A neural network framework for SoC (system on chip) is one developing point in the SI space. As of late, SoC has consolidated the AI procedure using Python. At the point when engineers consider an Intelligent framework on SoC, the coplan programming and equipment are huge. For this situation, little gadgets and implanted frameworks are the target framework. For this situation, one issue is the requirement for an immense memory and preparation power. Today, IoT gadgets are the powerhouse of the information age with their regularly expanding numbers and boundless infiltration.

Fig. 2–1 illustrates different application areas where SI is being used. Its implementation has been proven to be useful in areas ranging from communication, robotics, marketing, business analytics, and network optimization to autonomous vehicles and the health care sector. Also, swarm intelligence and machine learning arrangements are being coordinated to a wide range of applications, making items essentially increasingly "more intelligent." IoT equipment manufacturers should stay aware of the expanded throughput and proposal of new plans of action. Then again, AI/ML arrangements will create better outcomes.

2.1.1 Motivation

SI exhibits the collaborative behavior of social insects, such as the honeybee's dance, the wasp's building of a nest, the construction of a termite mound, or ants making a path. Interestingly, the algorithms used in SI are all about making a path with a colony of insects and the solution approach is calculated on a strange aspect of biology. Much scientific research has proved that individuals need not require any specific skill or knowledge to exhibit such complex behaviors.

The individual members in a social insects' colony are not informed before time regarding their local status or global objective. There is also no controller that guides them about their responsibility. The SI logic is equally distributed throughout all the agents who are fundamentally members to accomplish their respective objectives, but the individual agents can not complete their mission without cooperation from the rest of the agents. These interesting social insects exchange information such as the location of a food source or the possibility of any danger to their group friends. This type of communication between the individuals is based upon the concept of locality, where there is no idea about the global situation. This motivating approach of SI has solved many challenges when applied into technology. IoT is an emerging technology which is gradually dominating over the computing platforms very fast. This understanding motivated the authors to present an analytical study that will serve as a research platform for authors and researchers to perform further study in this direction.

2.1.2 Organization of the chapter

The chapter has been organized as follows. Section 2.1 introduces the chapter, Section 2.2 offers a detailed literature survey, Section 2.3 discusses the rational province of SI, and Section 2.4 presents the integration of IoT in developed and smart applications. Further, Section 2.5 presents SI applied in IoT systems, Section 2.6 illustrates SI using AI and ML in smart cities, and Section 2.7 concludes the chapter.

2.2 Literature review

In the above-discussed field of soft computing-based SI, the collaborative behavior of biologically inspired social insects, such as the honeybee's dance, the wasp's building of nest, the construction of termite mound, or ants making a path with a colony are measured as a strange aspect of methodological intelligence. Much scientific research has proven that individuals need not require any specific skill or knowledge to exhibit such complex behaviors. The individual members in a social insects' colony are not informed before time regarding their local status or global objective. There is also no controller that guides them about their responsibility. The SI is evenly distributed throughout all the agents who are basically members to achieve their respective goals, but the individual agents can not accomplish their mission without cooperation from the rest of the agents. Social insects exchange information and communicate the location of a food source or the existence of any danger to their kin. This type of communication between the individuals is based upon the concept of locality, where there is no idea about the global situation.

This section presents a review of literature research contributions based on SI in IoT and how performance of the system has clearly increased in complex situations. SI has an extraordinary ability in IoT-based frameworks to sensibly control their operations. The application of SI in the IoT has been acknowledged in numerous applications with explicit features for data calculation.

2.2.1 Swarm intelligence methods to optimize Internet of Things and network issues

SI calculations are entrenched and connected in surely understood intelligence-based issues where continuous activities are dealt with effectively. With IoT procedures being data concentrated, the nature of IoT administrations depend a lot on IoT data processing abilities.

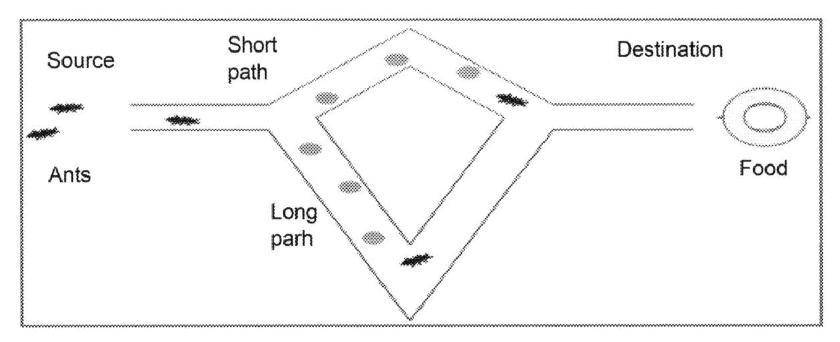

FIGURE 2–2 ACO mechanism.

With the assortment and volume of IoT data, SI calculations like ant colony optimization (ACO) can assume a significant job in improving the IoT forms. ACOs have been pioneered as powerful tools to explain order-based problems, for example, the traveling salesman problem and the quadratic assignment problem.

Fig. 2—2 shows the mechanism used in the ACO method where an ant selects a random path to reach its destination. In ACO, natural ants are capable of finding the shortest path from food sources to the destination without using a visual image. They are also capable of adapting to changes in the environment. There is a procedure in the particle swarm optimization (PSO) algorithm to acknowledge and organize congestion and its control as well as energy productive steering in the vehicle layer of Wireless Swarm Ad-hoc Networks (WSANs) [1]. The reproduction results demonstrate that the proposed PSO-based methodology gives a better execution regarding system lifetime and bundle drop proportion in contrast to the subterranean insect state advancement and the counterfeit honeybee province calculation.

2.2.2 PSOseed2 training neural network using velocity updation

For a modest memory and processing framework such as SoC, there is a proposition [2] that uses an improved molecule swarm advancement (PSO) calculation called the PSOseed2 calculation for preparing NN. The PSOseed2 calculation illuminates the untimely assembly of the standard PSO (SPSO) calculation by somewhat adjusting the speed update work without adding numerous computational errands to the SPSO calculation. We assessed this calculation on field programmable door cluster (FPGA)-based NN and programming-based NN and prepared these NNs with various PSO calculations that are SPSO, PSOseed, PSOseed2, and dissipative PSO. Exploratory outcomes with various datasets affirmed that the NNs prepared by the proposed PSOseed2 calculation would do better to acknowledgment rates and lower worldwide learning mistakes than the NN prepared by other PSO calculations.

The stability of a double-wheel self-supporting robot is maintained by following the transformed pendulum idea. The PWM strategy is adjusted for controlling the DC engines which are associated with the robot wheels. In the current robot models a tilt sensor is utilized, which makes them move in a straight line. Indeed, even with the interfacing encoder the tilt sensor does not reduce the errors. The proposed technique [3] utilizes both a gyroscope and an accelerometer to recognize tilt point and speeding up. Because of this the robot will most likely move in curvilinear way. Any errors from this sensor are overwhelmed by coupling a Kalman channel. By using IOT and a receiving port sending technique numerous robots can be controlled. The application areas of this sort of robot are moving wheel seats, hazard identification, and arranging short-range vehicles.

2.2.3 Human intranets and cluster-based energy optimization

IoT system [4] consists of three sorts of hubs: IoT hubs, central cluster nodes (CCNs), and base stations (BSs). The goal is to limit transmission between IoT hubs (IoTs), CCNs, and CCNs-BSs, and computational power (at CCNs), while fulfilling the necessities of the imparting hubs. The defined numerical model is a number programming issue. Three SI-based transformative calculations were proposed: (1) a discrete firecrackers calculation (DFWA), (2) a heap mindful DFWA (L-DFWA), and (3) a cross-breed of the L-DFWA and the low-intricacy biogeography-based advancement calculation to take care of the streamlining issue. The proposed calculations are populace-based metaheuristic calculations. They perform broad reenactments and measurable tests to demonstrate the presentation of the proposed calculations in contrast with the current ones.

2.2.4 Development of sparks explosion strategy

One of the current methods of the firecrackers-related algorithm is at present in trend and executing SI based logic, the presentation of which is dictated by much research work [5]. Researchers have developed an improved fireworks algorithm with landscape information for balancing exploration and exploitation.

2.2.5 Simulation of human brain storming

The advancement, usage, variation, and future bearings of another SI calculation—conceptualize streamlining (BSO) calculation—have been thoroughly overviewed. Conceptualizing improvement calculation is another promising SI calculation, which mimics the human conceptualizing process. Through the united activity and unique task, people in BSO are assembled and separated in the hunt for space/target space. Each person in the BSO calculation isn't just an answer for the issue to be streamlined, but also a data point to uncover the scene of the issue. In light of the improvements of conceptualizing enhancement calculations, various types of streamlining issues and certifiable applications could be solved [6].

2.2.6 Block chain-based data marketplace for trading

A bockchain-based, decentralized, and reliable data commercial center has been proposed [7], where IoT gadget sellers and AI (artificial intelligence) and ML (machine learning)-based arrangement suppliers may communicate and work together. By encouraging a straightforward data trade stage, access to data will be democratized and the assortment of administrations focusing on end-clients will increase. The proposed data commercial center is executed as a shrewd contract on the Ethereum blockchain and the swarm is utilized as the dispersed stockpiling stage.

2.2.7 Location tracking using wireless signal sampling

Among the numerous indoor positioning techniques is the sign example coordinating strategy, otherwise called the component unique mark strategy. Its working standard is through the gathering of remote sign examples to build up a unique finger impression database. When getting the restriction demand, the information data is put into the correlation with unique finger impression database to decide the definite area of the objective emanating similar remote sign examples. Numerous examinations depend on this guideline to create various calculations to improve the situating exactness. The deficiencies of this methodology is that the situating accuracy is effectively irritated by the loud condition conditions. These boisterous sign examples recorded in the unique finger impression database will lose the reference esteem, when low productivity ordering calculation or wrong models are utilized. A few novel AI-based calculations and ordering strategies for indoor situating precision improvements have been contemplated and broke down [8]. They are an altered swarm calculation, halfway least square (PLS) calculation, and hereditary calculations. Unique in relation to normal unique finger impression limitation strategies which utilize measurements-based models to portray the sign examples, such techniques utilize a few AI-based calculations to order the sign examples for restriction. Contrasted with different insights-based indoor unique mark limitation strategies, this technique can achieve an exactness of 95% with low advancement cost and a goal of 16 cm in an intricate PC research facility condition.

2.2.8 Secured shared data in Internet of Things environment

In IoT environment, the datasets are mutually shared [9] and such important analysis of data in the sterilized data set has been presented [10]. The multi-target molecule swarm streamlining structure and a calculation named as HCMPSO are used to adjust four symptoms.

A comprehensive outline in [11] gives an underlying comprehension of the specialized parts of SI calculations and their potential use in IoT-based applications. The current SI-based calculations are tried for thinking about their principle applications, at that point the current IoT-based frameworks are inspected that utilization SI-based calculations. At that point it features on patterns to unite SI and IoT-based frameworks. Existing powerful enhancement overviews center completely around developmental calculations and little on SI calculations. There is a necessity of a thorough overview committed to SI calculations to fill in the hole in the dynamic enhancement domain. In expansion to the standard subterranean insect state streamlining and molecule swarm improvement calculations;

ongoing SI applications to dynamic advancement issues should be considered. The current survey contains characterizations identified with both the algorithmic parts and the application issue.

An expansive audit on SI dynamic improvement intensified on a few classes of issues, for example, discrete, ceaseless, obliged, multigoal and arrangement issues, and certifiable applications have been displayed [12]. It centers around the improvement systems coordinated in SI calculations to address dynamic changes, the exhibition estimations and benchmark generators utilized in SIDO. At long last, a few contemplations about future bearings in the subject are given. Many research work deals with SI and Dynamic improvement, Ant settlement enhancement and Particle swarm advancement.

Real security issues for IoT with utilization of SI has been done [13]. It audits and arranges prevalent security issues with respect to the IoT layered design, notwithstanding conventions utilized for systems administration, correspondence, and the board. A blueprint of security prerequisites for IoT alongside the current assaults, dangers, and cutting edge arrangements has been displayed. Moreover, it organizes and maps IoT security issues against existing arrangements. More importantly, it examines how blockchain, which is the basic innovation for bitcoin, can be a key empowering agent to take care of numerous IoT security issues. It likewise distinguishes open research issues and difficulties for IoT security.

The above literature review section describes details of our literature study on the IoT and SI-based applications in different associated fields. In this review many related applications were studied and the challenges were explored. We comprehend that these challenges can be solved by proposing SI-based intelligent algorithms.

2.3 Rational province of swarm intelligence

SI is a term that has been coined from biological insect "swarms." It further refers to a loosely structured collection of interacting agents. Agents are individuals that belong to a group (but are not necessarily identical), they contribute to and benefit from the group, and they can recognize, communicate, and/or interact with each other. The instinctive perception of swarms is a group of agents in motion—but that does not always have to be the case. A swarm is better understood if thought of as agents exhibiting a collective behavior. Swarms be extended to other similar systems, some examples of swarms in nature are ant colony—agents: ants; flock of birds—agents: birds; traffic—agents: cars; crowd—agents: humans; immune system—agents: cells and molecules.

Fig. 2–3 shows Intelligent agents in SI. An agent plays a very important role in SI. It is a catalyst that can be viewed as perceiving its environment through sensors and acting upon that environment through effectors. Movements of rational agents depend on two factors. An ideal rational agent should do whatever action is expected to maximize its performance measure, on the basis of the evidence provided by the percept sequence and whatever built-in knowledge the agent has. Factors on which Rationality depends are Performance measure

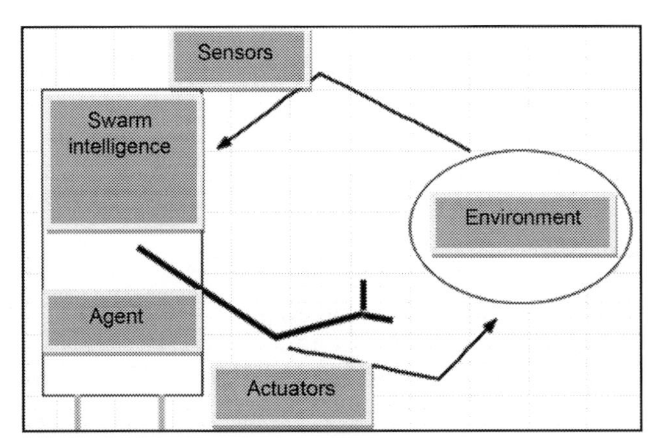

FIGURE 2–3 Intelligent agents in swarm intelligence.

(degree of success), Percept sequence (everything agent has perceived so far), Agents knowledge about the environment, and Actions that agent can perform. An ACO technique is a popular case of SI. It is the first ACO system, developed by Marco Dorgo in 1992, ants are searching for food. In the shorter path a greater amount of pheromone is left by ants. The probability of taking a route is directly proportional to the level of pheromone on that route. As more and more ants take the shorter path, the pheromone level increases. It efficiently solves problems like vehicle routing, network maintenance, and the traveling salesperson.

PSO is a method of SI application in which the population-based stochastic optimization procedure is used. It was built up and populated by Dr. Eberhart and Dr. Kennedy in 1995. In the techniques of PSO, the potential solutions, called particles, fly through the problem space by following the current optimum particles. It has been applied in many areas: function optimization, artificial neural network training, fuzzy system control etc. The most important application area of SI is robot swarms. Swarms provide the possibility of enhanced task performance, high reliability (fault tolerance), low unit complexity, and decreased cost over traditional robotic systems. They can accomplish some tasks that would be impossible for a single robot to achieve. Swarm robots can be applied to many fields, such as flexible manufacturing systems, spacecraft, inspection/maintenance, construction, agriculture, and medicine work. An operator assumes a significant job in SI. It is an impetus that can be seen as seeing its condition through sensors and following up on that condition through effectors.

An example of SI is the construction of termites where the worker insects modify their behavior as a result of which the nest is formed. Similarly business organizations also are formed by collaboration, interaction, and communication between individual workers. There interactions are communicated throughout the colony and therefore the colony can find solutions to its problems, whereas the individuals cannot solve the problems. These collective behaviors are termed as self-organizing behaviors. These theories can be used to demonstrate how social insects demonstrate complex collaborative behavior that come from communication from individual behavior.

Two commonly used SI algorithms are ACO and PSO. ACO refers to the study of artificial systems modeled after the behavior of real ant colonies and are useful in solving discrete optimization problems. It was originally called the ant system (AS) and it has been applied to the traveling salesman problem (and other shortest path problems) and several NP-hard Problems. It is a population-based metaheuristic used to find approximate solutions to difficult optimization problems. PSO refers to a population-based stochastic optimization technique. It searches for an optimal solution in the computable search space.

Fig. 2–4 shows different applications of PSO to solve smart city-based applications. The PSO concept gained inspiration from swarms of bees, flocks of birds, schools of fish, etc. In PSO individuals strive to improve themselves and often achieve this by observing and imitating their neighbors. Each PSO individual has the ability to remember. PSO has simple algorithms and low overheads. It is more popular in some circumstances than genetic/evolutionary algorithms. This method has only one operation calculation. There is a parameter called velocity which is a vector of numbers that are added to the position coordinates to move an individual.

The advantages of using SI are that the systems are scalable because the same control architecture can be applied to a couple of agents or thousands of agents. The systems are flexible because agents can be easily added or removed without influencing the structure. The systems are robust because agents are simple in design, the reliance on individual agents is small, and failure of a single agent has little impact on the system's performance. Using these methods the systems are able to adapt to new situations easily.

Recently, there have been multiple uses of SI techniques. The US military is applying SI techniques to control unmanned vehicles. NASA is applying SI techniques for planetary mapping. Medical research is trying SI-based controls for nanobots to fight cancer. SI techniques are applied to load balancing in telecommunication networks. In the entertainment industry these techniques are used for battle and crowd scenes.

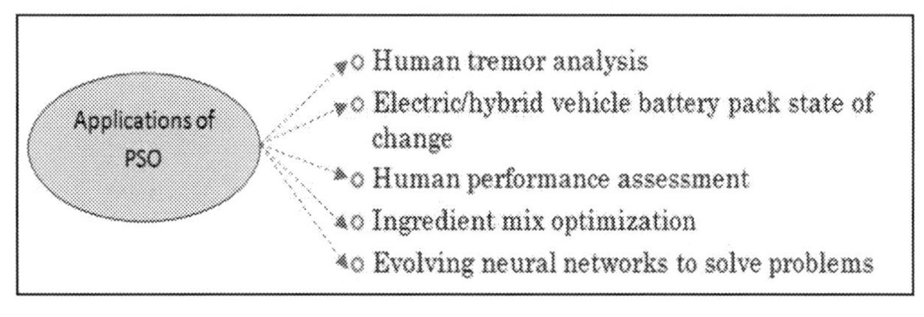

FIGURE 2–4 Applications of PSO.

2.4 Integration of Internet of Things in developed and smart applications

Basically, IoT is a system that comprises several gadgets, sensors that can speak with each other. The AI and ML capacities in the IoT help the system to handle the data received from associated devices. Further, this information is sent to the client or is utilized in choosing further activity, such as modifying the gadgets, and so forth. Urban communities that work toward improving the lives of their kin with the assistance of advanced innovations and enormous information are advanced urban areas [14].

Fig. 2–5 displays various applications of IoT in different sectors of society. There are numerous application fields of IoT from online shopping, flight service, and business analytics to wearable technology, smartphones, and smart home appliances. Pretty much every nation thinks that it is hard to deal with the developing deluge of residents in different urban city regions. Urban communities are being advanced by information sharing and the investigation of man-made consciousness, and obviously, a great many sensors. Metropolitan companies are embracing new advancements to spare operational expenses and augment the efficiencies of existing resources. An ongoing report by the US-based ABI Research noted that advanced cities can possibly save as much as US$5 trillion in annual expenses in the following 4 years. Notwithstanding, grasping an all-encompassing methodology is of great importance to guarantee the accomplishment of the tasks that are presently being executed [14]. Table 2–1 presents technical details of IoT and its applications in allied fields.

Table 2–1 describes details of study on IoT-based applications in different associated fields. In this review these applications were analyzed and the related functionalities were monitored. We comprehend that these challenges can be solved by using SI.

2.4.1 Integrated applications of Internet of Things with artificial intelligence and machine learning

It is a very important task to find solutions for many problems, every one of the issues in one go and solving all the problems which may not be conceivable within a short time period.

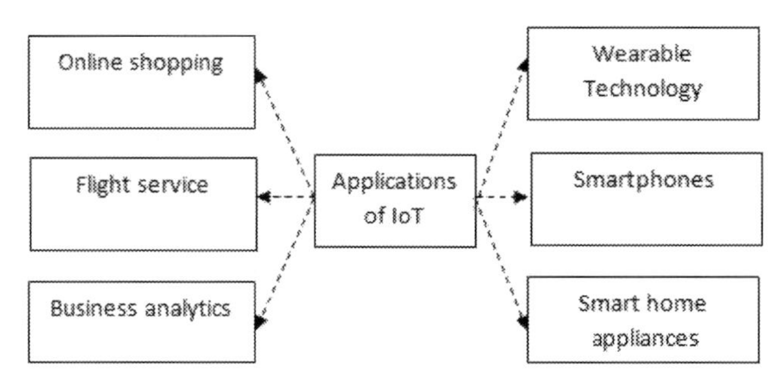

FIGURE 2–5 Displays various applications of IoT in different sectors of society.

Table 2–1 Details of Internet of Things applications in allied fields.

Literature authors	Research subject	Year	Associated fields
Frustaci et al. [15]	Evaluating critical protection issues of the Internet of Things world: present and future challenges	2018	Cloud computing, IoT, public administration, random forests, protection of data, smart cities
AboBakr and Azer [16]	Internet of Things ethics challenges and legal issues	2017	Computer network protection, IoT, social networking
Dalipi et al. [17]	EC-IoT: an easy configuration framework for constrained Internet of Things devices	2016	Embedded systems, IoT, machine-to-machine communication, protection of data
Chen et al. [18]	A vision of IoT: applications, challenges, and opportunities with China perspective	2014	Data privacy, IoT, socioeconomic effects
Yadav et al. [19]	IoT: challenges and issues in Indian perspective	2018	Internet, security, wireless networks
Giaffreda et al. [20]	A pragmatic approach to solving Internet of Things interoperability and protection problems in an eHealth context	2016	IoT, strategic planning
Park et al. [21]	Internet of Things routing architecture with autonomous systems of things	2014	Command and control systems, IoT, microcontrollers, military communication, wide area networks
Krishna and Gnanasekaran [22]	A systematic study of protection issues in Internet of Things (IoT)	2017	Cameras, IoT
Sezer [23]	T1C: Internet of Things Protection: Threats, Protection challenges and Internet of Things protection research and technology trends	2018	Data privacy, ethical aspects, IoT
Singh and Singh [24]	Internet of Things (IoT): protection challenges, business opportunities and reference architecture for E-commerce	2015	IoT, protection, E-commerce
Li et al. [25]	A survey on the development and challenges of the Internet of Things (IoT) in China	2018	Data privacy, ethical aspects, health care, IoT, medical information systems
Rhee [26]	Catalyzing the Internet of Things and smart cities: global city teams challenge	2016	Computer network protection, Internet, IoT, radiofrequency identification, wireless sensor networks
Jalaian et al. [27]	Evaluating LoRaWAN-based Internet of Things devices for the tactical military environment	2018	Internet, IoT, internetworking
Garg [28]	A lucid Internet of Things challenge to sustainable society	2018	Internet, IoT
Ray et al. [29]	Protection validation in Internet of Things space	2016	IoT, protocols
Venkatesan et al. [30]	Design of a smart gateway solution based on the exploration of specific challenges in IoT	2017	Groupware, handicapped aids, health care, hospitals, open systems, protection of data

(Continued)

Table 2–1 (Continued)

Literature authors	Research subject	Year	Associated fields
Shahid and Aneja [31]	Internet of Things: vision, application areas and research challenges	2017	IoT, smart cities
Alrashdi et al. [32]	AD-IoT: anomaly detection of Internet of Things cyberattacks in smart city using machine learning	2019	Data privacy, IoT, protection of data
Hong [33]	Challenges of name resolution service for information centric networking toward IoT	2016	Information dissemination, information retrieval, Internet, IoT, telecommunication network routing
Colley and Crabtree [34]	Object based media, the Internet of Things and databox	2018	Commerce, information technology, IoT, organizational aspects, retail data processing
Songqing Chen, Weisong Shi	Research challenges in edge computing	2015	Electronic commerce, IoT, protection of data
Dlaminiand Johnston [5]	The use, benefits and challenges of using the Internet of Things (IoT) in retail businesses: a literature review	2016	IoT, research and development
Biswas and Giaffreda [35]	Internet of Things and cloud convergence: opportunities and challenges	2014	IoT, telecommunication network routing
Routhand Pal [36]	A survey on technological, business and societal aspects of Internet of Things by Q3, 2017	2018	Cloud computing, IoT

As pointed out by Teradata's CTO [14], Stephen Brobst during his ongoing cooperation with columnists, the problem and purpose of each smart city can be extraordinary. A few territories may have the transportation framework as their essential concern, while others may be progressively concerned over the personal satisfaction of the general population. Brobst additionally proposed that city enterprises the world over should concentrate on improving proficiency and adopting a self-subsidizing strategy for undertakings as opposed to raising assessments.

2.4.2 Smart traffic management and road lights

Smart urban communities guarantee that their natives get from point A to point B as securely and effectively as could be expected under the circumstances. To accomplish this, regions go to IoT advancement and actualize smart traffic arrangements [37]. IoT-based smart urban communities make the upkeep and control of road lights progressively direct and smart. Outfitting streetlights with sensors and associating them to a cloud enables the adjustment of the lighting timetable to the lighting zone. Smart lighting arrangements accumulate information on illuminance, development of individuals and vehicles, and consolidate it with authentic and relevant information (e.g., unique occasions, open transport plan, time of day and year, and so on.) and break it down to improve the lighting plan. Subsequently, a smart lighting arrangement "tells" a streetlight to diminish, light up, turn on, or turn off the light dependent on the external conditions.

For example, when walkers cross the street, the lights around the intersections can change to a more splendid setting; when a transport is relied upon to stop at a transport stop, the streetlights around it may very well be naturally set to be brighter than those further away, and so on [37].

2.4.3 Management of waste products

This is certifiably not an exceptionally productive methodology since it prompts the ineffective utilization of waste holders and superfluous fuel utilization by waste gathering trucks. IoT-empowered smart city arrangements help to advance waste gathering plans by following waste dimensions, just as giving course enhancement and operational examination. Each waste holder gets a sensor that accumulates the information about the dimension of the loss in a compartment. When it is near a specific limit, the waste administration arrangement gets a sensor record, forms it, and sends a notice to a truck driver's versatile application. Along these lines, the truck driver exhausts a full holder, abstaining from discharging half-full ones.

Smart traffic arrangements utilize various kinds of sensors, that is, GPS information from drivers' smartphones to decide the number, area, and the speed of vehicles. In the meantime, smart traffic lights associated with a cloud the board stage permit observing green light timings and consequently adjust the lights dependent on current traffic circumstance to counteract blockage. Also, utilizing verifiable information, smart answers for traffic the board can anticipate where the traffic could proceed to take measures to avoid potential blockage.

For instance, being a standout amongst the most traffic-influenced urban communities on the planet, Los Angeles has executed a smart traffic answer to control traffic flow. Street surface sensors and closed-circuit TV cameras send constant updates about the traffic stream to a central traffic stage. The stage breaks down the information and advises the clients of jams and traffic sign breakdowns by means of work area client applications. Also, the city is sending a system of smart controllers to consequently make second-by-second traffic light modifications, responding to changing traffic conditions progressively.

With the assistance of GPS information from drivers' smartphones (or street surface sensors implanted in the ground on parking spaces), smart stopping arrangements decide if the parking spaces are being used or empty and make a continuous stopping map. At the point when the nearest parking space turns out to be free, drivers get a notice and utilize the guide on their telephone to discover a parking space quicker and simpler than indiscriminately driving around.

2.4.4 Smart transportation using Internet of Things

The information from IoT sensors can uncover examples of how residents use transport. Open transportation administrators can utilize this information to improve journeys, accomplishing a more elevated amount of security and timeliness. To do a progressively modern investigation, smart open transport arrangements can join numerous sources, for example, ticket deals and traffic data. In London, for example, some train administrators foresee the stacking of train traveler vehicles on their treks all through the city. They join the information from ticket deals, development sensors, and CCTV cameras introduced along the stage. Breaking down this information, train administrators can anticipate how every vehicle will load up with travelers. At the point when a train comes into a station, train administrators urge travelers to spread along the train to amplify the stacking.

2.4.5 Utilities of Internet of Things in smart city

IoT-prepared smart urban areas enable natives to set aside cash by giving them more command over their home utilities. IoT empowers various ways to deal with smart utilities:

1. Smart meters and charging: With a system of smart meters, regions can furnish residents with financially savvy availability to utilities organizations' IT frameworks. Presently, smart associated meters can send information legitimately to an open utility over a telecom arrange, furnishing it with solid meter readings. Smart metering permits utilities organizations to charge precisely for the measure of water, energy, and gas devoured by every family.
2. Revealing utilization designs: A system of smart meters empowers utilities organizations to increase more noteworthy perceivability and perceive how their clients expend energy and water. With a system of smart meters, utilities organizations can screen request continuously and divert assets as important or urge purchasers to utilize less energy or water on occasion of lack.
3. Remote checking: IoT smart city arrangements can likewise give natives utility administration administrations. These administrations enable residents to utilize their smart meters to track and control their utilization remotely. Moreover, if an issue (e.g., a water spillage) happens, utilities organizations can tell householders and send authorities to fix it.

2.4.6 Internet of Things-focused environment

IoT-driven smart city arrangements permit following parameters basic for a sound situation so as to keep up them at an ideal dimension. For instance, to screen water quality, a city can convey a system of sensors over the water lattice and associate them to a cloud the executives stage. Sensors measure pH level, the measure of disintegrated oxygen and broke up particles. On the off chance that spillage happens and the concoction structure of water changes, the cloud stage triggers a yield characterized by the clients. Another utilization case is observing air quality. For that, a system of sensors is sent along occupied streets and around plants. Sensors accumulate information on the measure of CO, nitrogen, and sulfur oxides, while the focal cloud stage examines and pictures sensor readings, with the goal that stage clients can see the guide of air quality and utilize this information to call attention to regions where air contamination is basic and work out proposals for residents.

2.4.7 Public safety

For improving open security, IoT-based smart city advances offer continuous checking, examination, and basic leadership instruments. Joining information from acoustic sensors and CCTV cameras throughout the city with the information from web-based live-feeds and breaking down it, open security arrangements can foresee potential wrongdoing. This will enable the police to stop potential culprits or effectively track them. For instance, in excess of 90 urban communities in the United States utilize a gunfire recognition arrangement. The arrangement uses associated mouthpieces introduced all through a city. The information from mouthpieces is sent to the cloud stage, which examines the sounds and recognizes a discharge. The stage estimates the time it took for the sound to reach the receiver and assesses the area of the firearm. At the point when the shot and its area are recognized, the cloud program notifies the police by means of a mobile application.

2.5 Swarm intelligence applied in Internet of Things systems

IoT-based frameworks are based on high computational logic and they are dynamic collections of elements (smart objects) which more often than not need decentralized control. SI frameworks are decentralized, self-sorted calculations used to determine complex issues with dynamic properties, deficient data, and constrained calculation capacities. SI calculations incorporate insect settlement improvement, molecule swarm advancement, honeybee enlivened calculations, bacterial searching streamlining, firefly calculations, fish swarm enhancement, and some more. These techniques have been demonstrated to be great strategies to address troublesome advancement issues under stationary conditions.

Fig. 2–6 depicts various applications of SI in a range of IoT-related applications. It includes some optimization problem, some health care applications, some network path finding and cost optimization, and some of the issues are related to blockchain. A blockchain is a new technology-based concept connected with security and cryptographic algorithms. In this technique, every block makes a link to its previous code using a hash tag function and related information. It ensures better security in network and IoT devices depending on the associated technology.

2.5.1 Particle swarm optimization-based spectrum genetic algorithm

Cognitive communication networks have developed as a dependable innovation to deal with a huge number of associated gadgets in the IoT. To accomplish this, range detecting undertaking ought to be trailed by ongoing tuning of transmission parameters with the goal that the targets of least transmit control, and most extreme throughput could be accomplished for various administration types. The basic leadership module for psychological radio is

Applications of swarm intelligence in IoT fields /frameworks	
Network congestion control and power efficiency in IoT routing	Cluster based IoT and energy optimization
Swarm intelligence methods to optimize IoT problems	Development of sparks explosion strategy
PSOseed2 Training Neural Network using velocity updation	Simulation of human brain storming
Control and balance of swarm robots using port forwarding	Block chain based data marketplace for trading
Human intranets with swarms	Location tracking using wireless signal sampling

FIGURE 2–6 Applications of swarm intelligence in various fields of IoT.

responsible to reach at some independent choice for a lot of transmission parameters as indicated by the transmission situation. A molecule swarm streamlining-based basic leadership module [38] has been intended to help three methods of activity. The recreation results have been contrasted and Real coded Genetic Algorithm (GA) that has diverse encoding mechanism when contrasted with generally predominant Binary coded Genetic Algorithm (BCGA) plot utilized previously. The outcomes show that the parameter adjustment for PSO-based motor outflanks the GA-based execution for all the transmission modes in CR-based IoTs.

2.5.2 Software-defined network for intelligent-Internet of Things

Several endeavors have been made to upgrade the incorporated observing based programming based networking idea of the huge scale smart Internet of things (I-IoT). Besides, the quantity of IoT gadgets in huge situations is expanding and an adaptable directing convention has hence turned out to be basic. Be that as it may, due to related asset confinements, truth be told, exceptionally little capacities can be arranged utilizing IoT hubs, chiefly those identified with the power supply. One answer for expanding system adaptability and drawing out the life of the system is to utilize the portable sink (MS). In any case, deciding the ideal arrangement of data social affair focuses (SDG), ideal way, planning the whole system with MS in a energy productive way and delaying the life of the system present colossal difficulties, especially in huge scale systems.

Energy effective directing convention dependent on man-made brainpower, that is, molecule swarm enhancement and hereditary calculation [39] has been presented, for huge scale I-IoT arranges under the SDN and cloud engineering. The essential reason is to endeavor cloud assets, for example, stockpiling and data-focus units by utilizing a unified SDN controller-based AI to compute: a heap adjusted table of bunches, an ideal SDG, and an ideal way for the MS (MSopath). In addition, the proposed new directing method will avert huge energy dissemination by the bunch head and by all hubs all in all by booking the entire system. Subsequently, the SDN controller basically balances energy utilization by the system during

Table 2–2 Details of research carried out on the application of SI in IoT.

Sl. no	Authors	Research association	Year
1	Bui and Jung [40]	ACO-based dynamic decision-making for connected vehicles in IoT system	2019
2	Lin et al. [10]	A sanitization approach to secure shared data in an IoT environment	2019
3	Li et al. [41]	fast incremental learning with swarm decision table and stochastic feature selection in an IoT extreme automation environment	2019
4	Usman et al. [42]	UAV reconnaissance using bio-inspired algorithms: joint PSO and penguin search optimization algorithm (PeSOA) attributes	2019
5	Ali et al. [4]	Optimizing the power using firework-based evolutionary algorithms for emerging IoT applications	2019
6	Chakraborty and Datta [43]	SIIoT: a shortest path estimation and obstacle avoidance system for autonomous cars	2018
7	Al-Janabi and Al-Raweshidy [39]	A centralized routing protocol with a scheduled mobile sink-based AI for large scale I-IoT	2018
8	Graff and Karnapke [44]	From centralized management of robot swarms to decentralized scheduling	2018
9	Özyilmaz et al. [7]	IDMoB: IoT Data Marketplace on Blockchain	2018
10	Akram et al. [45]	Energy efficient localization in wireless sensor networks using computational intelligence	2018
11	Rauniyar et al. [46]	A new distributed localization algorithm using social learning-based PSO for IoT	2018
12	Manshahia [1]	SI-based energy-efficient data delivery in WSAN to virtualize IoT in smart cities	2018
13	Shih and Liang [8]	The improvement of indoor localization precision through partial least square (PLS) and swarm (PSO) methods	2018
14	Batth et al. [47]	Internet of robotic things: driving intelligent robotics of future—concept, architecture, applications and technologies	2018
15	Kaur et al. [38]	PSO based multiobjective optimization for parameter adaptation in CR-based IoTs	2018

the steering development process as it considers both the SDG and the development of the MS. Recreation results exhibit the adequacy of the recommended model by improving the system life expectancy up to 54%, volume of data collected by the MS up to 93% and lessening the deferral of the MSopath by 61% in correlation with different methodologies.

Table 2–2 illustrates details of the types of research associating SI with IoT.

2.5.3 NP-hard problem of localization algorithm

Developing applications in the IoT will rely upon the precise area of thousands of sent sensors. In any case, exact limitation of sent sensors hubs is an old style improvement issue which falls under NP-hard class of issues. In this manner a research work [46] proposes another distributed confinement calculation utilizing social learning-based molecule swarm improvement for IoT. With this method of social learning PSO (SL-PSO), it expects to do exact limitation of conveyed sensor hubs and decrease the computational multifaceted nature which will further upgrade the lifetime of these asset obliged IoT sensor hubs. Broad recreations are completed to demonstrate the compelling exhibition of the SL-PSO calculation in precise limitation. Trial results delineate that SL-PSO calculation can build combination rate as well as fundamentally diminish normal confinement mistakes compared with conventional molecule swarm streamlining (PSO) and its different variations.

2.5.4 Electric energy consumption using pattern recognition

Particle swarm-based optimization technique, a computational type of intelligence method, is connected in a research facility condition to perceive presence of an example between the net energy utilization by electric loads in the structure and the surrounding temperature alongside the inhabitance condition of the structure; and utilize the distinguished example to anticipate energy utilization sooner rather than later. The electric loads under thought incorporate lighting and warming, ventilation and conditioning units with checking and control capacities utilizing IoT gadgets and advances. Having this expectation capacity is incredibly helpful to guarantee adequate energy is created to fulfill the needs of the electric burdens at all times. Thus this decreases energy wasted because of overabundance generation [48].

2.5.5 Decentralized scheduling of robot swarms

In the coming years IoT frameworks may include hundreds or even more work stations. Dealing with these stations ought to be done consequently, as individually writing computer programs is dull and tends to lead to mistakes. Consequently, a swarm scheduler [44] has been suggested that gets a determination of the activity to be done, and plans it on the best applicant robot. While this scheduler works fine for a predetermined number of hubs, it depends on a unified methodology and won't scale well when the quantity of hubs achieves hundreds or even thousands. The favorable circumstances and detriments of the concentrated methodology are additionally featured and two distinctive decentralized renditions are exhibited.

Another issue of smart traffic control is presented here showing how this scenario can be improved using very intelligent swarm technique. In the context of smart traffic control, a large number of nanorobots are coordinated in a decentralized and distributed manner at the junction points where an intelligent logic has to be applied regarding any critical decision in traffic points. Such controlling points include an estimation of the total number of vehicles, checking congestion status, arrival of new vehicles, decisions to be taken for rerouting the vehicle in other direction or to continue in the same route, what should be the best strategy to decide how many stop points there should be in congested traffic, controlling information regarding other traffic points, etc. The idea used here is that multiple simple nanorobots which act as microcontrollers at various points can undertake complex decision-making logic in a collaborative and distributed way in a group. In SI, the collaborative behavior of social insects, such as the honeybee's dance, the wasp's building of a nest, the construction of termite mounds, or the ants making a path with a colony, are measured as a strange aspect of biology.

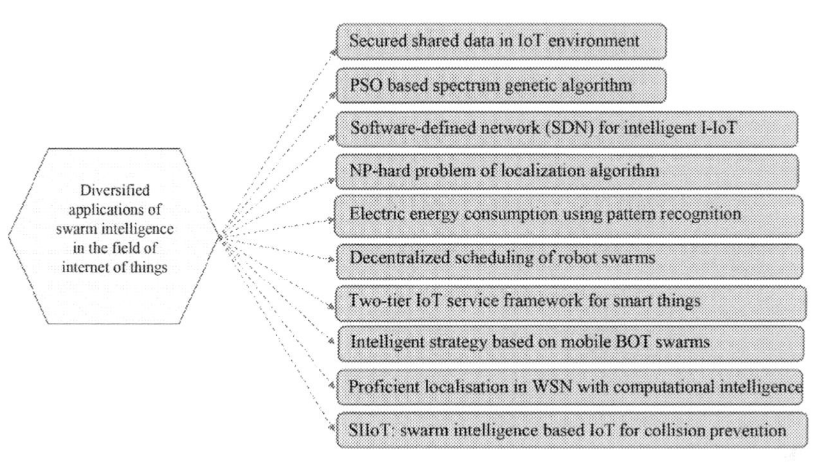

FIGURE 2–7 Diversified applications of swarm intelligence in IoT fields.

Many scientific researches have proved that individuals need not require any specific skill or knowledge to exhibit such complex behaviors. The individual members in a social insects colony are not informed before time regarding their local status or global objective. There is also no controller that guides them about their responsibility.

Fig. 2–7 demonstrates diversified applications of SI in IoT fields which includes energy consumption using pattern recognition, robot swarms and their correct movement, mobile BOT swarms, and localization problems in network using CI (computational intelligence) methods.

2.5.6 Two-tier Internet of Things service framework for smart things

Fig. 2–7 shows diverse applications of SI in IoT fields in current technology-based smart culture. Together with the tuning of IoT to help different modern applications, collaboration and coordination is a promising procedure for fulfilling prerequisites that are past the limit of any single thing. To address this test, a two-level IoT administration structure is proposed [49], where the functionalities given by smart things are embodied into IoT administrations, which are sorted into administration classes. The administration system is developed by considering the conjuring probability between administration classes, and administration class chains are produced utilizing conventional web administration structure methods, where the useful determination of specific prerequisites is considered [50,51]. Thinking about elements, for example, spatial and transient imperatives, energy proficiency, and the configurability of IoT administrations, choosing IoT administrations for the instantiation of administration classes contained in chains is decreased to a multiobjective and multiconstrained advancement issue. Heuristic calculations, for example, the genetic-based algorithm (GA), ACO, and PSO, are embraced to look for ideal IoT administration pieces. A test assessment demonstrates that PSO performs superior to the GA and ACO in scanning for ideal IoT

Table 2–3 Details of application challenges related to SI in IoT.

Sl. no	Authors	Year	Application challenges
1	Bui and Jung [40]	2019	Connected vehicles, IoT, vehicle dynamics, heuristic algorithms, decision-making, routing
2	Lin et al. [10]	2019	Security, data privacy, genetic algorithms, databases, particle swarm optimization, decision-making
3	Li et al. [41]	2019	Intelligent sensors, real-time systems, data mining, automation, edge computing, cloud computing, particle swarm optimization, stochastic processes
4	Usman et al. [42]	2019	Reconnaissance, fish, optimization, drones, particle swarm optimization, sociology
5	Ali et al. [4]	2019	Power optimization, evolutionary algorithm
6	Chakraborty and Datta [43]	2018	Cloud computing, collision avoidance, estimation, autonomous automobiles, automobiles, autonomous vehicles, heuristic algorithms
7	Al-Janabi and Al-Raweshidy [39]	2018	Artificial intelligence, sensors, routing protocols, cloud computing, wireless sensor networks, routing
8	Graff and Karnapke [44]	2018	Robots, trajectory, collision avoidance, schedules, complexity theory, dynamic scheduling
9	Özyilmaz et al. [7]	2018	Peer-to-peer computing, companies, monitoring, memory
10	Akram et al. [45]	2018	Wireless sensor networks, power demand, transceivers, power generation, sensors, wireless communication, task analysis
11	Rauniyar et al. [46]	2018	Optimization, wireless sensor networks, particle swarm optimization, IoT, routing, monitoring, distance measurement
12	Manshahia [1]	2018	SI, WSN
13	Shih and Liang [8]	2018	Databases, fingerprint recognition, machine learning algorithms, wireless communication, classification algorithms, statistics, mathematical model
14	Batth et al. [47]	2018	Robot sensing systems, robot kinematics, IoT, cloud computing, computer architecture
15	Kaur et al. [38]	2018	Modulation, genetic algorithms, sociology, statistics, optimization, linear programming, cognitive radio

WSN, Wireless sensor network.

administration creations and decreases the energy utilization, in this manner extending the systems lifetime.

Table 2–3 illustrates details of application challenges related to SI in IoT.

2.5.7 Intelligent mobile bot swarms

Biological quadcopters and their family are growing quickly, and honeybee estimated renditions are right now being pondered. An energizing vision for the IoT is a swarm of flying bots (automated honeybees) that can give a tremendous aggregate intelligence to social affairs information [52]. Be that as it may, this vision can't be executed yet on the grounds that the swarm requires new techniques for conveying capacity to the bots and new designs for amassing crude data into aggregate intelligence. As needs be, the useful frameworks of today will include a solitary bot working in an area. Expanding on those frameworks, it very well may be figured out how to make foundations to help swarms of bots that could support a city, state, or nation. It can likewise be foreseen that an incredible cooperative energy could be made between a progressively fit single bot and a swarm of less fit yet omnipresent bots.

2.5.8 Proficient localization in wireless sensor network with computational intelligence

Remote sensor networks comprise many detecting gadgets which are appropriated within a given zone. Every sensor hub comprises various heterogeneous parts, for example, control supply, CPU, memory, and a handset. Since the area of sensors is required in a large portion of the wireless sensor networks, trilateration-based localization has been utilized to find the sensors in the system. A parameter-investigation of the connected PSO variations is performed, prompting results that show algorithmic upgrades of up to 21% in the assessed goals.

2.6 Swarm intelligence using artificial intelligence and machine learning in smart cities

An important part of SI is influenced by AI and ML [53]. Algorithms used in SI are inspired by AI and ML and the sensors used in IoT use these algorithms to improve the proficiency of existing gadgets. AI and ML have likewise changed the manner in which individuals shop online nowadays. Indeed, even tech-firms like Smart Sight Innovations that offer eCommerce advancement arrangement have executed AI and AI capacities into shopping entrances.

In many IoT-based applications, the SI is evenly distributed throughout all the agents who are basically members to achieve their respective goals, but the individual agents can not accomplish their mission without cooperation from the rest of the agents. Social insects exchange information and communicate the location of a food source or the existence of any danger to their mates. This type of communication between the individuals is based upon the concept of locality, where there is no idea about the global situation. The implicit changes that takes place in the surroundings for communication are the key factor to be considered during the design of an intelligence-based algorithm in any soft computing-based strategy. Insects modify their activities because of the earlier changes made by their mates in the location. Another example of SI is the construction of termites where the worker insects modify their behavior and as a result the nest is formed. Similarly business organizations also are formed by collaboration, interaction, and communication between individual workers. Interactions are communicated throughout the colony and therefore the colony can find solution to the problems, whereas the individuals cannot solve the problems. These collective behaviors are termed as self-organizing behaviors. These theories can be used to demonstrate how social insects demonstrate complex collaborative behavior that arises from communications from individuals behavior [54].

The UN predicts that by 2050, the world's urban populace is probably going to have doubled and number 6.7 million individuals. As the quantity of urban occupants develops, urban communities face new chances and difficulties. To avert natural decay, keep away from sanitation issues, relieve traffic clog, and frustrate urban wrongdoing, districts can use the IoT.

IoT can possibly tame the weight of urbanization and make everyday living progressively more agreeable and secure. With the invention of smart homes, enthusiastic urban areas, and perceptive everything, the IoT has risen as a territory of amazing effect, potential, and development, with Cisco, Inc. forecasting there being 50 billion associated gadgets in 2020. Be that as it may, the vast majority of these IoT gadgets are anything but difficult to hack. Normally, these IoT gadgets are constrained by figure, stockpiling, and system limits, and in this way they are more powerless against assaults than other endpoint gadgets, for example, advanced cells, tablets, or PCs.

2.7 Conclusion

This chapter present an outline of the application of SI approaches in many advanced industrial and scientific uses of IoT. It has been shown that in all these cases the performance has improved up to a great extent in terms of better robotics/networks, shortest path determination from source to destination, reducing complexities in IoT devices and many more. Other than these, there are many autonomous vehicles and drones [43] which are the modern age associated autos and are getting expanded consideration from both the scholarly community and industry. Among a few open research headings in such vehicles, two most significant are the briefest way estimation and obstruction shirking framework. A SI, Cloudlet and IoT based way to deal with location both the difficulties has been given [43] proposition of an ACO calculation for the most brief course estimation and vector field based crash evasion framework for future self-ruling autos. The epic part of the examination is in coordinating the referenced systems into a Cloudlet based IoT engineering for the self-sufficient vehicles.

A broad review of the practiced methodologies in SI has been done in this chapter providing details of the methodologies. The very purpose of this review is to offer technical information about two magnificent technologies of the present time and to set up a basic research platform that joins and collaborates IoT with very intelligent SI.

References

[1] M.S. Manshahia, Swarm intelligence-based energy-efficient data delivery in WSAN to virtualise IoT in smart cities, IET Wirel. Sens. Syst. 8 (6) (2018) 256–259. Available from: https://doi.org/10.1049/iet-wss.2018.5143.

[2] Y. Hoshino, Hardware/software co-design SoC-system for a neural network trained by particle swarm optimization, in: 2017 IEEE 10th International Workshop on Computational Intelligence and Applications (IWCIA), Hiroshima, 2017, pp. 1–1.

[3] G. Chinnadurai, H. Ranganathan, Balancing and control of dual wheel swarm robots by using sensors and port forwarding router, in: Third International Conference on Advances in Electrical, Electronics, Information, Communication and Bio-Informatics, Chennai, 2017, pp. 532–536.

[4] H.M. Ali, W. Ejaz, D.C. Lee, I.M. Khater, Optimising the power using firework-based evolutionary algorithms for emerging IoT applications, IET Netw. 8 (1) (2019) 15–31.

[5] N.N. Dlamini, K. Johnston, The use, benefits and challenges of using the Internet of Things (IoT) in retail businesses: a literature review, in: 2016 International Conference on Advances in Computing and Communication Engineering (ICACCE), Durban, 2016, pp. 430–436.

[6] S. Cheng et al., A comprehensive survey of brain storm optimization algorithms, in: 2017 IEEE Congress on Evolutionary Computation (CEC), San Sebastian, 2017, pp. 1637–1644.

[7] K.R. Özyilmaz, M. Doğan, A. Yurdakul, IDMoB: IoT data marketplace on blockchain, in: 2018 Crypto Valley Conference on Blockchain Technology (CVCBT), Zug, 2018, pp. 11–19. <https://doi.org/10.1109/CVCBT.2018.00007>.

[8] C. Shih, C. Liang, The improvement of indoor localization precision through partial least square (PLS) and swarm (PSO) methods, in: 2018 IEEE Sensors Applications Symposium (SAS), Seoul, 2018, pp. 1–6.

[9] M. Frey, F. Große, M. Günes, Energy-aware ant routing in wireless multi-hop networks, in: 2014 IEEE International Conference on Communications (ICC), Sydney, NSW, 2014, pp. 190–196. <https://doi.org/10.1109/ICC.2014.6883317>.

[10] J.C. Lin, J.M. Wu, P. Fournier-Viger, Y. Djenouri, C. Chen, Y. Zhang, A sanitization approach to secure shared data in an IoT environment, IEEE Access. 7 (2019) 25359–25368.

[11] O. Zedadra, A. Guerrieri, N. Jouandeau, G. Spezzano, H. Seridi, G. Fortino, Swarm intelligence-based algorithms within IoT-based systems: a review, J. Parallel Distrib. Comput. 122 (2018) 173–187. ISSN 0743-7315. Available from: https://doi.org/10.1016/j.jpdc.2018.08.007.

[12] M. Mavrovouniotis, C. Li, S. Yang, A survey of swarm intelligence for dynamic optimization: algorithms and applications, Swarm Evolut. Comput. 33 (2017) 1–17. ISSN 2210-6502. Available from: https://doi.org/10.1016/j.swevo.2016.12.005.

[13] M.A. Khan, K. Salah, IoT security: review, blockchain solutions, and open challenges, Fut. Gener. Comput. Syst. 82 (2018) 395–411. ISSN 0167-739X. Available from: https://doi.org/10.1016/j.future.2017.11.022.

[14] <https://dzone.com/articles/> (accessed 1.06.19).

[15] M. Frustaci, P. Pace, G. Aloi, G. Fortino, Evaluating critical security issues of the IoT world: present and future challenges, IEEE Internet Things J. 5 (4) (2018) 2483–2495.

[16] A. AboBakr, M.A. Azer, IoT ethics challenges and legal issues, in: 2017 12th International Conference on Computer Engineering and Systems (ICCES), Cairo, 2017, pp. 233–237.

[17] E. Dalipi, F. Van den Abeele, I. Ishaq, I. Moerman, J. Hoebeke, EC-IoT: an easy configuration framework for constrained IoT devices, in: 2016 IEEE 3rd World Forum on Internet of Things (WF-IoT), Reston, VA, 2016, pp. 159–164.

[18] S. Chen, H. Xu, D. Liu, B. Hu, H. Wang, A vision of IoT: applications, challenges, and opportunities with China perspective, IEEE Internet Things J. 1 (4) (2014) 349–359.

[19] E.P. Yadav, E.A. Mittal, D.H. Yadav, IoT: challenges and issues in Indian perspective, in: 2018 3rd International Conference on Internet of Things: Smart Innovation and Usages (IoT-SIU), Bhimtal, 2018, pp. 1–5.

[20] R. Giaffreda, L. Capra, F. Antonelli, A pragmatic approach to solving IoT interoperability and security problems in an eHealth context, in: 2016 IEEE 3rd World Forum on Internet of Things (WF-IoT), Reston, VA, 2016, pp. 547–552.

[21] S. Park, N. Crespi, H. Park, S. Kim, IoT routing architecture with autonomous systems of things, in: 2014 IEEE World Forum on Internet of Things (WF-IoT), Seoul, 2014, pp. 442–445.

[22] B.V.S. Krishna, T. Gnanasekaran, A systematic study of security issues in Internet-of-Things (IoT), in: 2017 International Conference on I-SMAC (IoT in Social, Mobile, Analytics and Cloud) (I-SMAC), Palladam, 2017, pp. 107–111.

[23] S. Sezer, T1C: IoT security: threats, security challenges and IoT security research and technology trends, in: 2018 31st IEEE International System-on-Chip Conference (SOCC), Arlington, VA, 2018, pp. 1–2.

[24] S. Singh, N. Singh, Internet of Things (IoT): security challenges, business opportunities & reference architecture for E-commerce, in: 2015 International Conference on Green Computing and Internet of Things (ICGCIoT), Noida, 2015, pp. 1577–1581.

[25] Y. Li, Y. Guo, S. Chen, A survey on the development and challenges of the Internet of Things (IoT) in China, in: 2018 International Symposium in Sensing and Instrumentation in IoT Era (ISSI), Shanghai, 2018, pp. 1–5.

[26] S. Rhee, Catalyzing the Internet of Things and smart cities: global city teams challenge, in: 2016 1st International Workshop on Science of Smart City Operations and Platforms Engineering (SCOPE) in Partnership with Global City Teams Challenge (GCTC) (SCOPE-GCTC), Vienna, 2016, pp. 1–4.

[27] B. Jalaian, T. Gregory, N. Suri, S. Russell, L. Sadler, M. Lee, Evaluating LoRaWAN-based IoT devices for the tactical military environment, in: 2018 IEEE 4th World Forum on Internet of Things (WF-IoT), Singapore, 2018, pp. 124–128.

[28] A. Garg, A lucid IoT challenge to sustainable society, in: 2018 Second International Conference on Inventive Communication and Computational Technologies (ICICCT), Coimbatore, 2018, pp. 1529–1533.

[29] S. Ray, S. Bhunia, Y. Jin, M. Tehranipoor, Security validation in IoT space, in: 2016 IEEE 34th VLSI Test Symposium (VTS), Las Vegas, NV, 2016, p. 1.

[30] V.P. Venkatesan, C.P. Devi, M. Sivaranjani, Design of a smart gateway solution based on the exploration of specific challenges in IoT, in: 2017 International Conference on I-SMAC (IoT in Social, Mobile, Analytics and Cloud) (I-SMAC), Palladam, 2017, pp. 22–31.

[31] N. Shahid, S. Aneja, Internet of Things: vision, application areas and research challenges, in: 2017 International Conference on I-SMAC (IoT in Social, Mobile, Analytics and Cloud) (I-SMAC), Palladam, 2017, pp. 583–587.

[32] I. Alrashdi, A. Alqazzaz, E. Aloufi, R. Alharthi, M. Zohdy, H. Ming, AD-IoT: anomaly detection of IoT cyberattacks in smart city using machine learning, in: 2019 IEEE 9th Annual Computing and Communication Workshop and Conference (CCWC), Las Vegas, NV, 2019, pp. 305–310.

[33] J. Hong, Challenges of name resolution service for information centric networking toward IoT, in: 2016 International Conference on Information and Communication Technology Convergence (ICTC), Jeju, 2016, pp. 1085–1087.

[34] J.A. Colley, A. Crabtree, Object based media, the IoT and Databox, in: Living in the Internet of Things: Cybersecurity of the IoT – 2018, London, 2018, pp. 1–6.

[35] R. Biswas, R. Giaffreda, IoT and cloud convergence: opportunities and challenges, in: 2014 IEEE World Forum on Internet of Things (WF-IoT), Seoul, 2014, pp. 375–376.

[36] K. Routh, T. Pal, A survey on technological, business and societal aspects of Internet of Things by Q3, 2017, in: 2018 3rd International Conference On Internet of Things: Smart Innovation and Usages (IoT-SIU), Bhimtal, 2018, pp. 1–4.

[37] <https://www.scnsoft.com/blog/iot-for-smart-city-use-cases-approaches-outcomes> (accessed 1.06.19).

[38] A. Kaur, A. Kaur, S. Sharma, PSO based multiobjective optimization for parameter adaptation in CR based IoTs, in: 2018 4th International Conference on Computational Intelligence & Communication Technology (CICT), Ghaziabad, 2018, pp. 1–7.

[39] T.A. Al-Janabi, H.S. Al-Raweshidy, A centralized routing protocol with a scheduled mobile sink-based AI for large scale I-IoT, IEEE Sens. J. 18 (24) (2018) 10248–10261.

[40] K.N. Bui, J.J. Jung, ACO-based dynamic decision making for connected vehicles in IoT system, IEEE Trans. Ind. Inform. 15 (10) (2019) 5648–5655. Available from: https://doi.org/10.1109/TII.2019.2906886.

[41] T. Li, S. Fong, R.C. Millham, J. Fiaidhi, S. Mohammed, J. Fiaidhi, Fast incremental learning with swarm decision table and stochastic feature selection in an IoT extreme automation environment, IT Professional 21 (2) (2019) 14–26.

[42] M.R. Usman, M.A. Usman, M.A. Yaq, S.Y. Shin, UAV reconnaissance using bio-inspired algorithms: joint PSO and Penguin Search Optimization Algorithm (PeSOA) attributes, in: 2019 16th IEEE Annual Consumer Communications & Networking Conference (CCNC), Las Vegas, NV, 2019, pp. 1–6.

[43] T. Chakraborty, S.K. Datta, SIIoT: a shortest path estimation and obstacle avoidance system for autonomous cars, in: 2018 Global Internet of Things Summit (GIoTS), Bilbao, 2018, pp. 1–6. <https://doi.org/10.1109/GIOTS.2018.8534532>.

[44] D. Graff, R. Karnapke, From centralized management of robot swarms to decentralized scheduling, in: 2018 IEEE International Conference on Internet of Things and Intelligence System (IOTAIS), Bali, 2018, pp. 141–147.

[45] J. Akram, Z. Najam, H. Rizwi, Energy efficient localization in wireless sensor networks using computational intelligence, in: 2018 15th International Conference on Smart Cities: Improving Quality of Life Using ICT & IoT (HONET-ICT), Islamabad, 2018, pp. 78–82.

[46] A. Rauniyar, P. Engelstad, J. Moen, A new distributed localization algorithm using social learning based particle swarm optimization for Internet of Things, in: 2018 IEEE 87th Vehicular Technology Conference (VTC Spring), Porto, 2018, pp. 1–7.

[47] R.S. Batth, A. Nayyar, A. Nagpal, Internet of robotic things: driving intelligent robotics of future – concept, architecture, applications and technologies, in: 2018 4th International Conference on Computing Sciences (ICCS), Jalandhar, 2018, pp. 151–160.

[48] G. Bedi, G.K. Venayagamoorthy, R. Singh, Pattern recognition for electric energy consumption prediction in a laboratory environment, in: 2017 IEEE Symposium Series on Computational Intelligence, Honolulu, HI, 2017, pp. 1–8.

[49] M. Sun, Z. Shi, S. Chen, Z. Zhou, Y. Duan, Energy-efficient composition of configurable Internet of Things services, IEEE Access. 5 (2017) 25609–25622. Available from: https://doi.org/10.1109/ACCESS.2017.2768544.

[50] M. Rath, B. Pati, B.K. Pattanayak, Design and development of secured framework for efficient routing in vehicular Ad Hoc network, Int. J. Bus. Data Commun. Netw. 15 (2) (2019) 55–72. Available from: https://doi.org/10.4018/IJBDCN.2019070104.

[51] M. Rath, B. Pati, Security assertion of IoT devices using cloud of things perception, Int. J. Interdiscip. Telecommun. Netw. 11 (4) (2019) 17–31. Available from: https://doi.org/10.4018/IJITN.201910010.

[52] P.C. Salmon, P.L. Meissner, Mobile bot swarms: they're closer than you might think!, IEEE Consum. Electron. Mag. 4 (1) (2015) 58–65. Available from: https://doi.org/10.1109/MCE.2014.2360057.

[53] A. Ritthipakdee, A. Thammano, Primate swarm algorithm for continuous optimization problems, in: 2017 18th IEEE/ACIS International Conference on Software Engineering, Artificial Intelligence, Networking and Parallel/Distributed Computing (SNPD), Kanazawa, 2017, pp. 11–15.

[54] L. Rosenberg, M. Lungren, S. Halabi, G. Willcox, D. Baltaxe, M. Lyons, Artificial swarm intelligence employed to amplify diagnostic accuracy in radiology, in: 2018 IEEE 9th Annual Information Technology, Electronics and Mobile Communication Conference (IEMCON), Vancouver, BC, 2018, pp. 1186–1191.

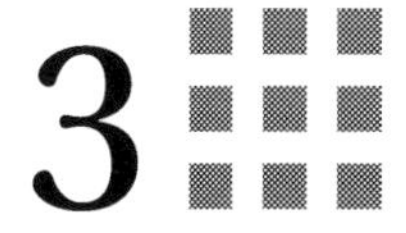

Arbitrary walk with minimum length based route identification scheme in graph structure for opportunistic wireless sensor network

S. Sivabalan, S. Dhamodharavadhani, R. Rathipriya

DEPARTMENT OF COMPUTER SCIENCE, PERIYAR UNIVERSITY, SALEM, INDIA

3.1 Introduction

During recent years the generation of new mobile devices with larger communication areas without the consideration of physical transportation has been emerging. To manage this problem some networks have been created such as pocket switched networks (PSNs) and but face a huge challenge with larger mobility. As a solution, these networks require an opportunistic forwarding algorithm that has been proposed in earlier work to solve the end-to-end delay, high packet delivery ratio (PDR), and the message delivery ratio in space and time. The example of opportunistic forwarding algorithms has been planned to provide promising wireless networking applications [1−3]. The major challenge of the present opportunistic forwarding schema is that it doesn't need to satisfy the following conditions such as less end-to-end delay, higher PDR, and less communication cost to reach source to destination.

The fundamental mobility procedure argues for some of the solutions to solve these problems, but it is not efficient for opportunistic networks, since the individual walks greatly influence the network performance [4,5]. Analyzing the intrinsic characteristics of the user in the network is achieved through complex network analysis [6]. So the analysis of the social network with opportunistic forwarding algorithms has been of great interest in recent years. In this work we mainly focus on the analysis of opportunistic data forwarding schema for social networks.

Recently, some work has been undertaken in social metrics to devise the opportunistic forwarding assessment example in Fig. 3−1. SimBet [7], Bubble [8], and People Rank [9] are the most recently used methods. All three methods are primarily based on two aspect points of view from society networks:

1. Community with the closest association tends to live in communities; and
2. Populace within a community might contain different popularity.

Swarm Intelligence for Resource Management in Internet of Things. DOI: https://doi.org/10.1016/B978-0-12-818287-1.00006-1

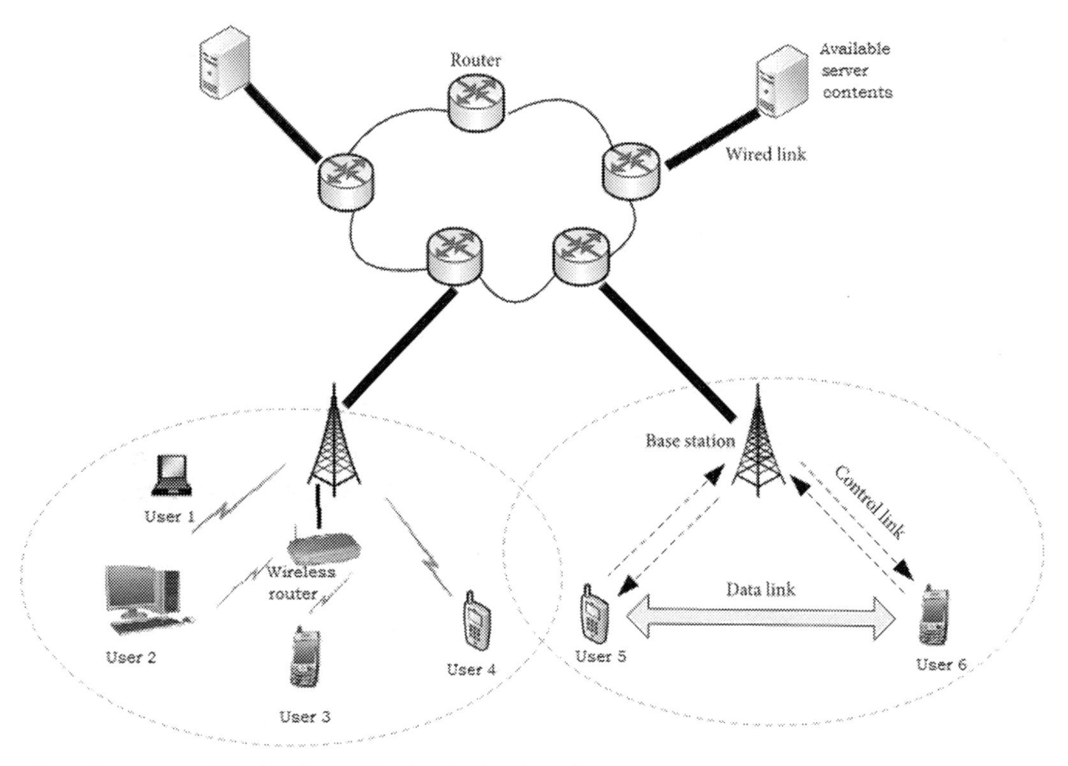

FIGURE 3–1 Opportunistic data forwarding for wired and wireless scenario.

As such, the ever more "popular" nodes are almost certainly selected as a carrier to communicate messages among disconnected communities [10], until a node corresponds to the similar society by way to reach the destination [9]. Community is a basic conception in methodical sociology. Everybody in public has a communal location which demonstrates his/her communal individuality and preserves his social individuality [11]. Centrality is one more way to show one's communal position as it reflects influence in a group [12]. Tie-strength is inclined to estimate the social graph at an atomic stage. It focuses on the strength of association for a dyad [13], which symbolizes a couple of nodes and the edge connecting them.

However, none of the *opportunistic network* work uses the following three major global parameters to estimate the value of nodes exploiting ego networks [14], between centrality [15] and PageRank [16] metrics, the nodes from one to each other are ranked in the network. But the major problem with these methods is that they don't count the frequency value of the reply nodes, which are sent from the source to destination nodes in the network. In order to overcome this problem, we have added the mean and standard deviation value. The mean and standard deviation metrics estimate the performance of opportunistic routing. Incorporating social metrics into opportunistic routing is the preliminary effort of the best of our information. The overlapped community formation is also additionally measured to differentiate the association nodes from other nodes to reach a destination. Decayed aggregation graph (DAG) [17] technology measures the relationship among nodes in the community structure. Three real existing opportunistic networking scenarios are compared and experimented for a real-lifetime dataset. The proposed random walk-based opportunistic forwarding efficiency with mean and standard deviation (RWOFMSD) is outperforming the earlier methods, particularly in terms of mean delivery delay (MDD) and cost.

3.2 Proposed network model methodology

This chapter presents a novel RWOFMSD, based on this mean and standard deviation updating values in the DAG by applying the preprocessing methods. This results in the detection of the nodes as strong nodes, bridge nodes, and noise nodes in a disseminated manner, which works more appropriately. To perform this data forwarding schema, nodes in the network are symbolized as a DAG: $G=(V,E)$, where V denotes the set of nodes and E denotes the set of edges. Adjacency matrix is indicated through $W(t)=(w_{uv}(t))_{n\times n}$ and contact series are indicated through $N_{uv}(t)=(\mathrm{on_i},\mathrm{off_i})\mathrm{i}=1,2,\ \ldots,\ \mathrm{N}$ among nodes during time duration t, where the starting instant and ending instant of the youth contact is represented through tuple $(\mathrm{on}_i,\mathrm{off}_i)$, respectively, and total number of contacts are specified as N. Strength of association among nodes $w_{uv}(t)$ is calculated by the estimation of the decayed sum at any current time T in

$$w_{uv}(T)=\sum_{i=1}^{N} f(i)g^{(T-\mathrm{off}_i)} \tag{3.1}$$

ith contract period is specified through $f(i)=\mathrm{off}_i-\mathrm{on}_i$ and decayed function are specified as,

$$g(T-\mathrm{off}_i)\cdot g(T-\mathrm{off}_i)=e(-\times(T-\mathrm{off}_i)) \tag{3.2}$$

Exponential decay [18] is followed here. The space complications of DAG are examined to reformulate

$$w_{uv}(T)=\sum_{i=1}^{N}(\mathrm{off}_i-\mathrm{on}_i)\mathrm{e}^{-\beta(T-\mathrm{off}_i)} \tag{3.3}$$

For higher storage capacity and to reduce the complexity Eq. (3.3) is rewritten to continuous interval [0, T]:

$$w_{uv}(T)=\sum_{t\le T} h(t)\mathrm{e}^{-\beta(T-t)} \tag{3.4}$$

$$h(t)=\mathrm{off}_i-\mathrm{on}_i$$

The RWOFMSD the algorithm is presented here.

The proposed DAG structure consists of two major steps to characterize the opportunistic network in a graphical way by proposing Algorithms 3.1 and 3.2. In the first part of the work, the data forwarding schema (Fig. 3–2), nodes in the network are symbolized as a DAG, $G = (V, E)$, where V denotes the set of nodes and E denotes the set of edges considered as input. The total number of messages sent from source node to destination nodes in the opportunistic network is calculated as $l_s = \log n$ with frequency level of reply (RPLY) messages in the R arbitrary walks. Finally, compute the mean and standard deviation, $<$l; mean; stdDeviation$>$ values to RPLY frequency count message on the opportunistic network.

Algorithm 3.2 makes a decision as to whether the selected nodes are noise nodes all the way through the calculation of the length values $l = l_0$ from Algorithm 3.1 and thus the results for nodes are $stdDeviation * \alpha(\alpha = 20)$. Then consider the corresponding node as a noise node or else it becomes a strong node or bridge node, and repeat this process until all nodes in the DAG re completed.

The algorithm divides the noise user node region into a smaller noise region for the exact identification of noise nodes in the DAG using the network simulation tool. If the values of the noise node in the DAG S_1 are smaller than the noise region, then it is identified as an honest node or else it is considered as a noise node. This is applied for all numbers of nodes in the DAG.

The system is applied to the entire network through the random walk length estimation method in Algorithm 3.3 (Figs. 3–3 and 3–4) and it is applied to Algorithm 3.4. Initiate the algorithm with R random walks for each numbered node in the DAG and c the calculation of

Algorithm 3.1 Preprocessing (*G*, *h*)

1. $J = \text{h}$
2. For $i = 1$ to f do
3. Perform a random walk with length $l_s = \log n$ originating from source to destination nodes in the graph structure h
4. $J = J \cup$ the ending node of the random walk
5. End for
3. $l = l_{min}$
7. while $l <= l_{max}$ do
8. for $i = J.first() to J.last()$do
9. Perform R random walks with length l originating from node I
10. Get n_i as the number of nodes with frequency no smaller than t
11. End for
12. Output $< l, mean(\text{n}_\text{i}\text{:i}\epsilon\text{J}), stdDeviation(\text{n}_\text{i}\text{:i}\epsilon\text{J}) >$
13. $l = l + 100$
14. end while

Algorithm 3.2

1. $l = l_0$
2. while $l <= l_{max}$ do
3. Perform R random walks with length l originating from u
4. $M =$ the number of nodes whose frequency is no smaller than t
5. Let the tuple corresponding to length l in the outputs of Algorithm 3.1 be $< l, mean, stdDeviation >$
3. If $mean - m > stdDeviation * \alpha$ then
7. Output u is noise nodes
8. End the algorithm
9. End if
10. $l = l * 2$
11. end while
12. output u is strong nodes, bridge nodes

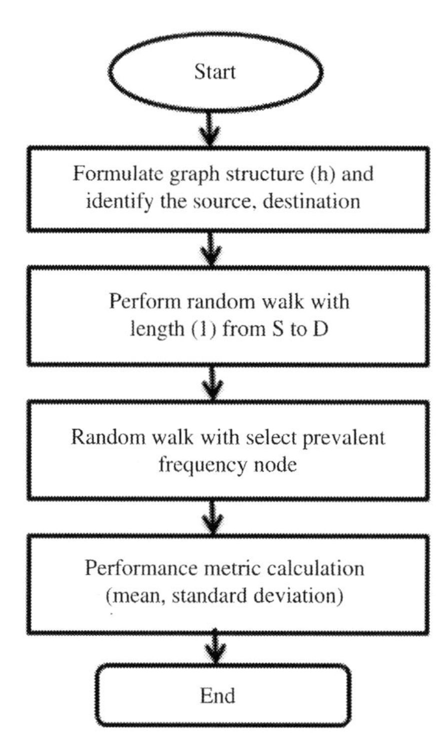

FIGURE 3–2 Flow of preprocessing of random walk theory based routing.

Algorithm 3.3 Walk length estimation (G, s)

1. $l = l_0/2$
2. $deadWalkRatio = 0$
3. while dead $WalkRatio < \beta do$
4. $l = l * 2$
5. dead $WalkNum = 0$
6. for $i = 1 toR$ do
7. Perform a incomplete arbitrary walk originating from s with length l
8. If the partial random walk is dead before it reaches l hops then
9. $DeadWalkNum++$
10. End if
11. End for
12. $DeadWalkRatio = deadWalkNum/R$
13. End while
14. Output l

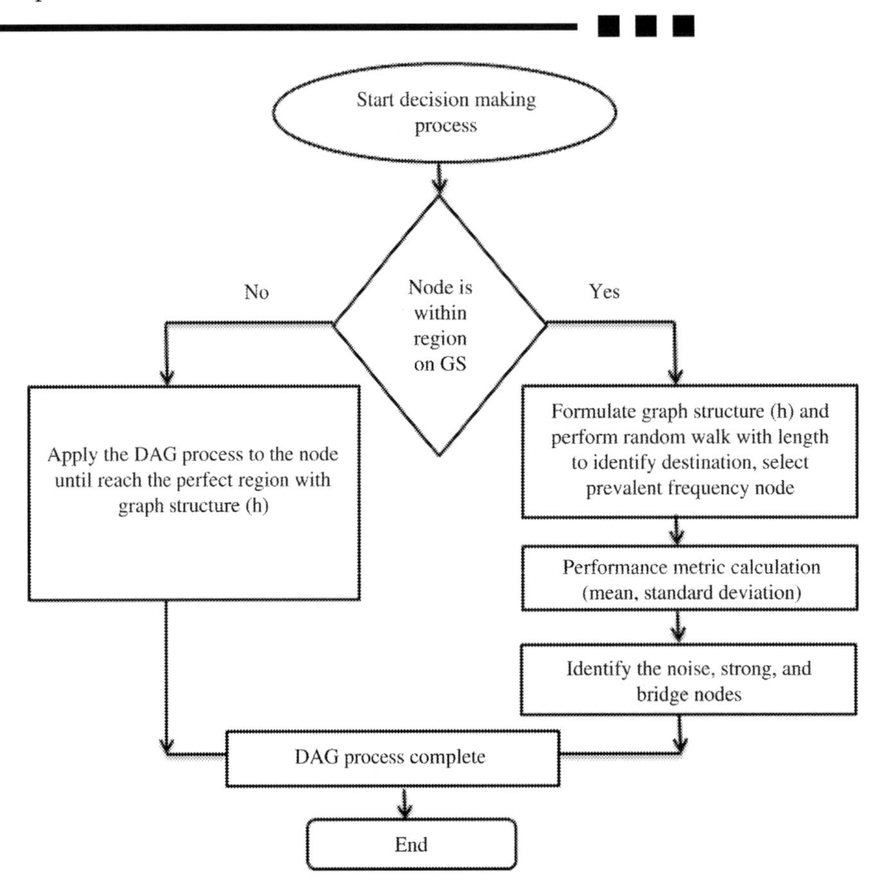

FIGURE 3–3 Flow of decision-making in DAG to identify a noise, strong, or bridge node.

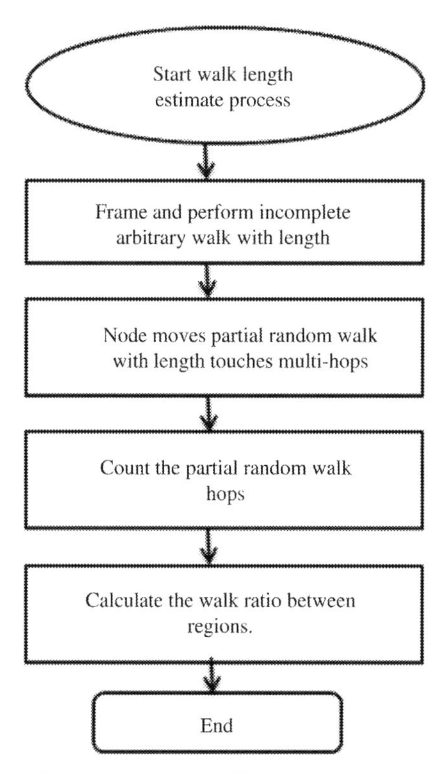

FIGURE 3–4 Flow of the estimation of arbitrary random walk length.

the life ratio walks. If it is smaller than the β it is considered as the noise node. Repeat this process until the maximum number of the honest nodes are identified.

Algorithm 3.4 detects the noise and honest node data forwarding opportunistics based on the random walk theory in a partial manner. It may not produce better results in all DAG, but in order to overcome these problems, in Algorithm 3.5 the noise node and honest node are directly identified based on the conductance evaluate. It performs sorting methods to sort the frequency count value of nodes for RPLY messages in the DAG theory, and concurrently inserts new nodes to identify noise nodes regions and honest nodes in the network presents in Fig. 3–5.

Finally, find the honest nodes among the deployed nodes in the network environment area. Fig. 3–6 displays the flow of Algorithm 3.5.

3.2.1 Message scheduling and buffer management

After the nodes are detected as honest nodes, noise nodes, and bridge nodes, the messages are transmitted from one node to another node with less communication cost. When the defenses attain their capability and new messages necessitate a buffer, communication is

Algorithm 3.4 Noise node detection ($G, S, IfromAlg3.$)

1. Set the frequency of all the nodes to be 0
2. For $I = 1toR$ do
3. Perform a incomplete arbitrary walk originating from node s with length l
4. $S.frequency++$
5. For $j = 1tol$ do
6. Let the jth hop of the partial random walk be node k
7. $K.frequecy++$
8. End for
9. End for
10. Traversed list = Sort the traversed nodes by their frequency in decreasing order
11. Counter = 0
12. $S = \phi$
13. Do
14. Counter = conductance(s)
15. For I = traversedList. first () to traversed List. last()do
16. If node $i \epsilon S$ then
17. Continue
18. If conductance ($\{i\} \cup S$)$<=$conductance(s) then
19. $S = \{i\} \cup S$
20. While (counter > conductance(S))
21. Output S

abandoned to make a decision using the buffer management algorithm. The messages in the communication are ordered based on the subsequent priority rules.

- The messages determination is broadcast primarily if they satisfy $I_d == I_i$ while intracommunity communication through larger priority value.
- The intercommunity communication is designed for messages with the intention that don't assure the following condition $I_d == I_i$.

In order to manage the buffer replacement algorithm, it needs to maintain the following principles.

- In a social network community, the user sends messages with less than buffer density value. The older message would be restored while the density value should be equivalent.
- The messages with $I_d == I_s$ are measured subsequently and messages with the lowest density will be outmoded.

Based on the scheduling and buffer management of the nodes will be classified in the following manner.

Algorithm 3.5 Combo (G, u, tuplesfromAlg.1)

1. $l = l_0$
2. while $l <= l_{max}$do
3. Perform R arbitrary walks through length l originating from u
4. m = the number of nodes whose frequency is no smaller than t
5. Let the tuple corresponding to length l in the outputs of Algorithm 3.1 be $< l, mean, stdDeviation >$
6. If mean $- m > stdDeviation * \alpha$then
7. Output u is Sybil
8. Traversed list = Sort the traversed nodes by their frequency in decreasing order
9. Counter = 0
10. for i = traversed List.first () to traversed List.Last() do
11. if node $i \in S$ then
12. continue
13. if conductance$(\{i\} \cup S) <= conductance(S)$ then
14. $S = \{i\} \cup S$
15. While (counter $> conductance(S)$)
16. Output S
17. End the algorithm
18. End if
19. $l = l * 2$
20. end while
21. output u is honest nodes and bridging nodes
22. $S = \phi$
23. Do
24. Counter = conductance(S)

- Node v is a noise node: The message m to node v does not promote designed for node u.
- Node u is a noise node, but node v is a strong or bridging node: Buffer removes messages m when node u a head m to node v.
- Neither u nor v is a noise node: Higher conductance metrics are larger if message m does not distribute the community of the target which goes to node u as noise node.

Evaluating higher conductance: Nodes with higher conductance are calculated earlier, but it is not practically appropriate to us an unidentified amount of neighbor's nodes and susceptible end-to-end pathway in social networks. In order to solve this problem principal component analysis (PCA) [17] technology has been developed to distinguish the overlapped community formation.

Determining k, the number of communities: in PCA if the value becomes larger the calculation of eigenvalues becomes more confusing, to solve this problem the top k eigenvectors are selected in the graph that satisfies Eq. (3.5),

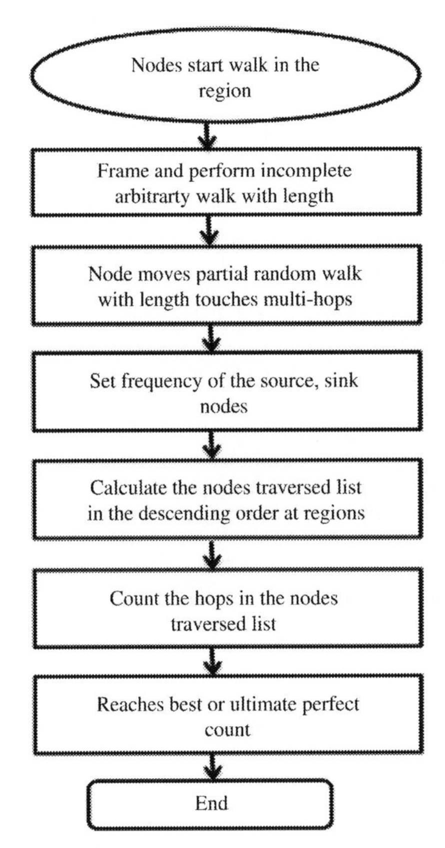

FIGURE 3–5 Noise node detection in partial manner.

$$\frac{\sum_{i=1}^{k} \lambda_i}{\sum_{i=1}^{k} \lambda_j} \geq R \tag{3.5}$$

and the ratio R generally belongs to the period [0.7, 0.9].

Identifying the noise nodes: identifying the noise nodes in the community PCA splits the nodes in two ways:

1. The principal components P_k, and
2. The opposed $= (x_{k+1}, \ldots, x_n)$.

and consequently, split the row vector α_u through $\alpha_u^{1,k}(\alpha_{u1}, \ldots, \alpha_{uk})$ and $\alpha_u^{k+1,n}(\alpha_{u,k+1}, \ldots, \alpha_{un})$ as nonnoise and noise nodes.

Determining the initial elements for each community: the preliminary node of the communal i is preferred as a node of u, s.t. $max\ |\alpha_{ui}|\ (u = 1, 2, \ldots, n)$ for eigenvector values in PCA x_i, and set $m_i = \alpha_{ui}$.

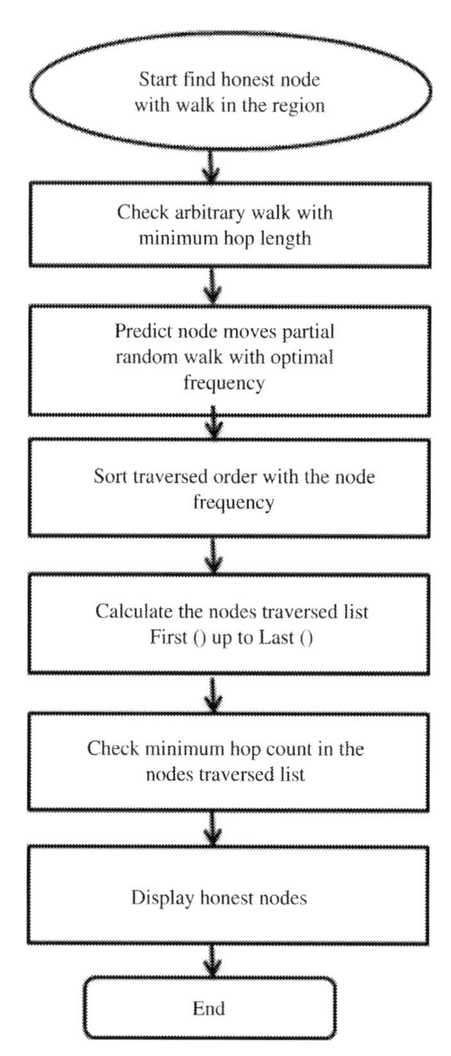

FIGURE 3–6 Find out honest nodes.

Termination condition of k-means: assume that every one of the nonnoise nodes is clustered, and the messages result in m_i being simplified in Eq. (3.6).

$$m_i = \frac{\left(\sum_{u \subset C_i} \alpha_u\right)}{n_i} \tag{3.6}$$

where n_i is the number of nodes corresponding to C_i. K means clustering results are specified by reducing the sum of the squared errors in Eq. (3.7),

$$J = \sum_{i=1}^{k} \sum_{u \subset C_i} (\alpha_u - m_i)^2 \tag{3.7}$$

3.2.2 Detecting the overlapped community structure

The overlapped community is detected based on the calculation of the two categories:

- Strong nodes
- Bridging nodes

The nodes belonging to the same community are termed strong nodes or else are named as bridging nodes. The subsequent steps are used to perform the relationship between the strong nodes and bridging nodes.

1. *Gateway (formally clustering) nodes*: The distance among one node and randomly selected nodes is defined as $dist(\alpha_u, m_i)$, and whether it is performed for community as s.t. $mindist(\alpha_u, m_i)(i = 1, 2, \ldots, k)$ in Eq. (3.8) where,

$$dist(\alpha_u, m_i) = \theta(u, i) = arcros \frac{\alpha_u m_i^T}{||\alpha_u||_2 ||m_i||_2} \tag{3.8}$$

   where (u, i) denotes the angle between α_u and m_i.
2. *Adjusting the categories of nodes*: The indistinct tag of nodes and unambiguous community formation are detected simultaneously after completion of step (1).

3.3 Experimentation results

In order to evaluate the performance of RWOFMSD with existing methods Prophet [22] and Direct Contact [20], open-loop fractional path loss compensational (OFPC) algorithms in terms of the MDD, cost and PDR. A rich variety of surroundings is discovering in the direction of cover and examines these results. NS2 simulation for produce the results and It can easily deploy nodes with different levels such that single, grid, horizontal, vertical nodes in with or without the cluster.

3.3.1 Mean delivery delay

Diverse message time to live results are shown in Fig. 3–7, to demonstrate the accuracy of MDD and furthermore clearly show that RWOFMSD speeds up the messaging more than the existing methods/algorithms mentioned above.

3.3.2 Cost

Fig. 3–8 RWOFMSD give the best results in terms of cost. The RWOFMSD method reduces the packet overhead ratio for the quick propagation process of mean delivery ratio better than the existing state-of-the-art message algorithms.

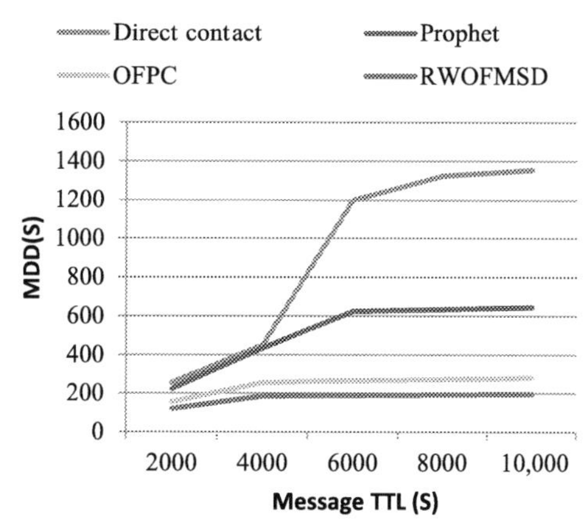

FIGURE 3–7 Mean delivery delay within different message time to live (TTL) for the State Fair.

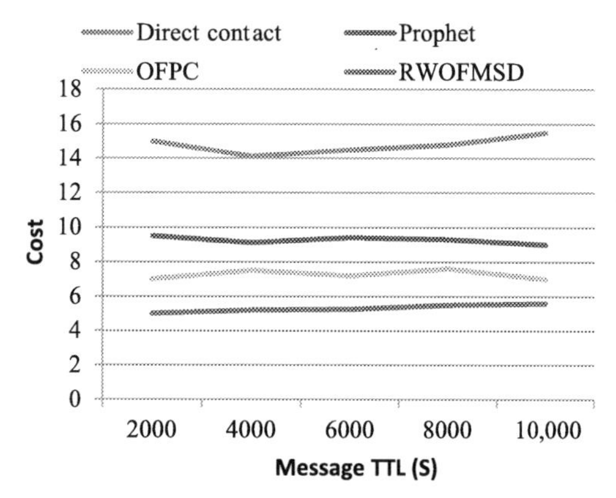

FIGURE 3–8 Cost of the average number of nodes infected by a message within a different message for the State FAIR.

3.3.3 Packet delivery ratio

PDR is shown in Fig. 3–9. RWOFMSD provides the best performance, attaining a higher PDR than the Bubble and OFPC methods.

3.4 Conclusion and future work

In this paper a novel opportunistic data forwarding scheme is designed to make the final decision for data forwarding or not for a selected path or route in the opportunistic network

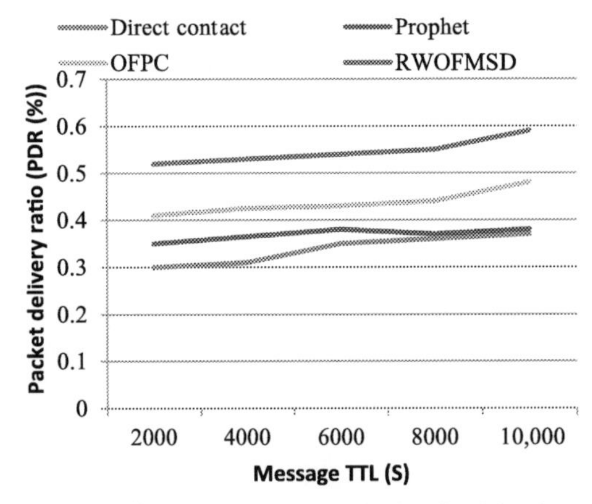

FIGURE 3–9 Packet delivery ratio within different message time to live (TTL) for the State Fair.

based on the calculation of the standard deviation and mean value for calculation of the frequency counter value by proposing a random walk theory. The opportunistic network is represented in the form of DAG. By calculation of mean and standard deviation values the proposed data forwarding methods detect the noise nodes during the community-based detection mechanism for single and numerous communities, rejecting the noised detected nodes and identifying honest nodes in the DAG structure. Then the results of the community detection conductance value are measured by proposing PCA and an overlapped community structure is also detected. Finally, the effectiveness of the proposed RWOFMSD scheme is validated in terms of MDD, PDR, and cost against the existing state-of-the-art. In our future work, we will apply the present RWOFMSD to measure the adaptive activities of real-time social network community and mobility activities in terms of contextual knowledge.

Acknowledgment

The first author gratefully acknowledges financial support from *UGC-RGNF* (Rajiv Gandhi National Fellowship) and its award letter-number of UGC No: F1-17.1/2014-15/RGNF-2014-15-SC-TAM-85083 (SA-III/website) dated: February 26, 2015. The second author acknowledges the UGC-Special Assistance Programme (SAP) for the financial support to her research under the UGC-SAP at the level of DRS-II (Ref. No. F. 5-6/2018/DRS-II (SAP-II)), July 26, 2018 in the Department of Computer Science.

Authors profile

Mr. S. Sivabalan pursed Bachelor of Computer Applications from Government Arts College, Salem, Tamil Nadu, India in year 2010, Master of Computer Application from Sona College of Technology, Salem, Tamil Nadu, India in year 2013 and Master of Philosophy in Computer Science from Periyar University, Salem, Tamil Nadu, India in year 2014. He is currently pursuing Ph.D. in Department of Computer Science, Periyar University, Salem, Tamil Nadu, India since 2015 as *UGC-RGNF* SRF Candidate. He has published research papers in reputed international journals and conferences including IEEE, Springer and it's also available online. Author main research work focuses on Mobile ad hoc Networks, Wireless Sensor Networks, Home Area Networks and Swarm, Non-Swarm based Computational Intelligence. He has 6 years of Research Experience.

Mrs. S. Dhamodharavadhani pursed Bachelor of Science, Master of Computer Application and Master of Philosophy in Computer Science from Mahendra Arts and Science College, Kalippatti, Tamil Nadu, India in year 2014. She is currently pursuing Ph.D. in Department of Computer Science, Periyar University, Salem, Tamil Nadu, India since 2015. She has published research papers in reputed international journals and conferences including IEEE and it's also available online. Her main research work focuses on Climate Data Analysis, Bio-inspired Computing, Big Data Analytics, Data Mining and Computational Intelligence. She has 5 years of Research Experience.

Dr. R. Rathipriya pursed Master of Computer Science in year 2003, Master of Philosophy in year 2004, Master of Computer Application in year 2008 and Ph.D. in Department of Computer Science, Bharathiar University, Coimbatore 2013, Tamil Nadu, India. She has published more than 42 research papers in reputed international journals including Thomson Reuters (SCOPUS & Web of Science) and conferences including IEEE, Springer and it's also available online. Her main research work focuses on Web Mining, Bioinformatics, Bio-inspired Optimization, Agro-Climate Data Analysis, Big Data Analytics, Data Mining and Computational Intelligence. She has 13 years of Teaching Experience and she has 10 years of Research Experience.

References

[1] A. Lindgren, A. Doria, O. Schelen, Probabilistic routing in intermittently connected networks, ACM SIGMOBILE Mob. Comput. Commun. Rev. 7 (3) (2003) 19–20.

[2] A. Mtibaa, M. May, C. Diot, et al., People rank: social opportunistic forwarding, in: IEEE Proc. Infocom, 2010, pp. 111–115.

[3] A.W. Wolfe, Social network analysis: methods and applications, Am. Ethnologist 24 (1) (1997) 219–220.

[4] D. Karaboga, An idea based on honey bee swarm for numerical optimization, Technical Report-TR06, Erciyes University, Engineering Faculty, Computer Engineering Department, Turkey, 2005.

[5] E. Cohen, M.J. Strauss, Maintaining time-decaying stream aggregates, J. Algorithms 59 (1) (2006) 19–36.

[6] E. Daly, M. Haahr, Social network analysis for routing in disconnected delay-tolerant MANETs, in: Proc. MobiHoc ACM, 2007, pp. 32–40.

[7] G.M. Bianco, Getting inspired from bees to perform large scale visual precise navigation, in: Proceedings of 2004 IEEE/RSJ International Conference on Intelligent Robots and Systems, Sendai, Japan, 2004, pp. 619–624.

[8] H.S. Sadeg, S. Yahi, Cooperative bees swarm for solving the maximum weighted satisfiability problem, in: IWAAN International Work Conference on Artificial and Natural Neural Networks, Barcelona, Spain, 2005, pp. 318–325.

[9] I. Rhee, M. Shin, S. Hong, et al., On the levy-walk nature of human mobility, in: Proc. IEEE Infocom, 2008, pp. 924–932.

[10] K. Fall, S. Farrell, DTN: an architectural retrospective, IEEE J. Sel. Areas Commun. 26 (5) (2008) 828–836.

[11] K. Xu, V. Li, J. Chung, Exploring centrality for message forwarding in opportunistic networks, in: IEEE, Wireless Communications and Networking Conference (WCNC), April 2010, pp. 1–6.

[12] L.D. Jiang, Q.Y. Zou, Z.Y. Cao, A queen-bee evolution based on genetic algorithm for economic power dispatch, in: UPEC 2004 39th International Universities Power Engineering Conference, Bristol, UK, 2004, pp. 453–456.

[13] M. Anupama, B. Sathyanarayana, Survey of cluster based routing protocols in mobile ad hoc networks, Int. J. Comput. Theory Eng. 3 (6) (2011) 806–815.

[14] M. Conti, M. Kumar, Opportunities in opportunistic computing, Computer 43 (1) (2010) 42–50.

[15] P. Hui, J. Crowcroft, E. Yoneki, BUBBLE rap: social-based forwarding in delay tolerant networks, in: Proc. MobiHoc ACM, 2008, pp. 241–250.

[16] R. Agarwal, M. Motwani, Survey of clustering algorithms for MANET, Int. J. Comput. Sci. Eng. 1 (2) (2009) 98–104.

[17] R.C. Shah, S. Roy, S. Jain, Data MULEs: modeling and analysis of a three-tier architecture for sparse sensor networks, in: Proc. 1st IEEE on Ad Hoc Networks, 2003, pp. 215–233.

[18] S. Okasha, Altruism, group selection and correlated interaction, Br. J. Philos. Sci. 56 (4) (2005) 703–725.

[19] S.A. Ade, P.A. Tijare, Performance comparison of AODV, DSDV, OLSR and DSR routing protocols in mobile ad hoc networks, Int. J. Inform. Technol. Knowl. Manag. 2 (2010) 545–548.

[20] R.C. Shah, S. Roy, S. Jain, Data MULEs: modeling and analysis of a three-tier architecture for sparse sensor networks, Ad Hoc Netw. 1 (2003) 215–233.

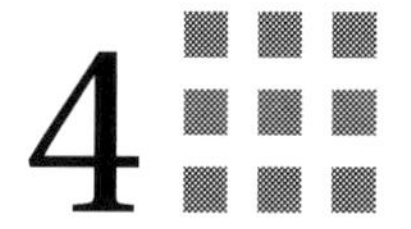

Cyberphysical systems in the smart city: challenges and future trends for strategic research

Mazen Juma, Khaled Shaalan

FACULTY OF ENGINEERING AND INFORMATION TECHNOLOGY, BRITISH UNIVERSITY IN DUBAI, DUBAI, UAE

4.1 Introduction

The world is becoming a digitally enabled network of individuals, objects, products, and services. Technology will be part of everything in the digital business of the future; the average person will live in a digitally enhanced world, wherein the lines between what is real and what is digital are obsolete. Productive digital services will provide everything, and intelligence will be embedded in everything [1].

Cyberphysical systems (CPSs) are intelligently networked systems with sensors, processors, and actuators installed within them that identify and interact with real-world aspects and human end-users. CPS supports real-time and ensures quality and extent of performance within safety-oriented applications, especially in the smart city model. In CPS systems, the cumulative behavior of the mixed physical and cyber aspects of the system consists of control, critical computing, identification, and networking embedded within all components. The acts of these elements and systems have to be synergized with each other and operate safely [2]. Amidst the technological revolution regarding the smart city, five major dominant trends will drive digital life through at least 2030.

These trends include artificial intelligence (AI) and advanced machine learning that is now at a significant changing process and continues to embolden hypothetically all technology-enabled services, things, and applications, creating smart systems capable of learning, adapting, and potentially acting on their own accord instead of adhering to preset instruction protocols [3]. Secondly, smart apps, including virtual assistants are a second trend, possessing the potential to overhaul the workplace by rendering daily tasks simpler, and consequently making users more efficient. However, such apps extend beyond digital assistants; all software types, like security tooling and applications within enterprises, for example, enterprise resource planning, are infused with AI-enabled capabilities [4].

The third trend is the new intelligent things, spanning three categories: autonomous vehicles, robots, and drones which experience continuous evolution thus impacting a more considerable

Swarm Intelligence for Resource Management in Internet of Things. DOI: https://doi.org/10.1016/B978-0-12-818287-1.00008-5

portion of the market, inviting a new era of digital business. The fourth trend is the virtual and augmented reality, transforming the way people interact together and how software systems construct an immersive setting. For example, virtual and augmented reality is designing training systems and remote experiences, which enable real and virtual worlds to overlap. Businesses can use this to animate graphics onto or within real-world objects [5].

Finally, digital twins, replicas of physical assets, will be capable of illustrating billions of things; a digital twin acts as an adaptive software replica of a system or physical object. This model emulates how physics interact with the object and how the object interacts with the environment, in addition to data supplied by planted sensors in the real world. Digital twins are capable of analyzing and modeling real-world parameters, act in response to changes, tweak operations, and add value [6].

The engineering research field of CPS continues to draw interest from governments, industries, and academia due to its influencing merits on society, the environment, and the economy. The new approach jumpstarted the creation of CPSs derived from analog computation. Ziegler-Nichols' tuning rules can be considered as a means to tweak the entire CPS to implement the needed behavior [7]. Technology has evolved various fields, which increased the need to build large-scale systems to provide for the booming society-related needs in a set of relatively scarce resources. It provides a reason, and a motive, for greater interest in research into the core of CPS [8].

Among recent technologies, CPS is an ever-growing concept representing the integration of computational and physical capabilities, which have variable applications, in the process, energy, and traffic control, medical devices, aviation, advanced automated systems, and smart structures. Businesses and manufacturers vigorously collaborate with telecommunication services suppliers, service providers, software producers, and even each other to create a confluence of their knowledge and competencies; these capabilities devise and operate cross-industry product innovation [9].

Recently, Big Data became a buzzword, said, and used by everyone. The concept had been known in the field of data mining since human-generated content flooded social networks; some tend to call it the web 2.0 as a concept from the end of 2004. Research organizations and communities devote themselves to this new research subject matter. CPSs provide solutions to unique challenges troubling our society.

These challenges are relevant to issues facing numerous industries and fields of application. CPS aid organizations to optimize, therefore to save cost, energy, and time. In the case of individuals, CPS deliver higher levels of comfort. For example, this can be applied in mobility assistance, networked safety, assisted living for senior citizens, and individual health care [10].

4.2 The state-of-the-art

CPS are sophisticated, multidisciplinary, next-generation, physically comprehending, and engineered systems in the smart city that implements computing solutions, the cyber part, into physical occurrence by employing transformative research approaches. This integration can be translated into observation, communication, and control elements of physical systems from the multidisciplinary view [11].

In less than five decades, cities around the world tripled from 548 back in 1970 to 1692 in 2016. Currently, over 54% of the global population is located in cities; this percentage may reach 66% by 2050. With roughly four billion people residing in cities nowadays, a ripple of global digital-era urbanization is at hand [9].

Growing portions of urban assortments, varying in size, are fast developing their smart cities. This global movement affects policy innovations, extensive worldwide investments in new technology applications, and data utilization methodologies, focusing on solving unstoppable urban growth and societal issues [12]. Smart cities' definition relies on capacity, political agenda, stakeholders, and the city's vision [13].

However, standard definitions share a common key aspect of "smartness" in an urban context: it is the use of communication and information technology solutions as core enablers of "smart" transformation of the city. Smarter cities are made possible today thanks to Internet connectivity available everywhere, social acceptance of technology, availability of sophisticated advanced data analytics, and possible large-scale interoperability of interconnected "things" [14].

Cities with resources, technological infrastructure, political leadership, and vision utilize digital technology as a facilitator and a vital infrastructural enabler that solves the challenges of population growth, urbanization, and financial and environmental issues.

Therefore a broad definition of a smart city would be an urban construct that blends information and communication technologies, enhancing livability, improving workability, maximizing sustainability, and improving the practices of governance, urban planning, and management. Alternatively, a data-focused definition defines it as a city capable of producing, harnessing, and analyzing data in a manner that allows for "intelligent" decisions and proactive and predictive analysis for an improved form of planning and improvement [15].

The UN considers smart cities, universally, as innovative cities that combine information and communication technologies to improve the quality of life. Additionally, a smart city works on increasing the efficiency of civil services, operations, and competitiveness while ensuring that it meets the needs of citizens, present and future, whether economic, social, or environmental. According to this concept, the crucial development for a sustainable smart city takes place in aspects such as lifestyle, services and infrastructure, intelligence and information, communications, people and society, environment and sustainability, administration and governance, finance and economy, and transportation and mobility [2].

Technically, the presence of infrastructure of energy and water-smart grids, digital governance capacity, urban systems of mobility, data improvement, and public inclusion initiatives, and buildings and structures are crucial to a smart city. A smart city project is a substantial technical and urban social transformation that revolutionizes the way people live in and interact with the city, data core enablers, and digital technology [16]. CPS, as a term, first appeared at the National Science Foundation in the United States around 2006. It ushered in the future of networking and information technology by 2007 [17].

The scientific community perceived it in different lights. Silva et al. [12] describe CPS as physical and engineered systems employing tracked, planned, controlled, and integrated operations by a computing and communicating core. Wang et al. [18] describe CPSs as the integration of computation with physical processes. Zhang et al. [19] describe them as systems embedded together and with their surrounding physical environment. Sivarajah et al. [10] describe them as biological, physical, and engineered systems with integrated, monitored, and controlled operations, done so via computational core. These components are interconnected at every scale, and the computing is profoundly implemented in every physical component, to find its way into materials [20]. The computational core can be defined as an embedded system that fetches real-time response; most of the time, it is distributed. The evolution of digital electronics has led to a significant surge in the number of systems that converge digital, or cyber, systems with the physical world. This process had been known as the CPS. CPS requires a significant amount of reasoning concerning unique challenges and sophisticated functions, reliability, and performance parameters [12].

Some studies focused on the underlying problem formulations, system-level requirements, and challenges in CPS. Xu and Duan [8] introduced the CPS concept and offered research pointers for CPS design. Tokody and Schuster [11] correctly pointed out the failure of standard abstraction layers, the necessity of positive timing behavior, and deficit in temporal semantics of current programming language models for CPS. Jeschke et al. [20] addressed system-level aspects of CPS by tackling scientific and social impact standpoints.

Liu et al. [21] proposed two approaches: one of them is cyberizing the physical while the other is physicalizing the cyber, both acting as means for blending the cyber aspect with the physical aspect in systems. Several existing surveys illustrate the comprehensive view of CPS. Stankovic [22] outlined CPS features, applications, and challenges, generally without in-depth details. Serpanos [23] described CPS research directions, specifications, and design with a shallow explanation of CPS applications and system-level requirements. Varghese and Buyya [17] touched on CPS characteristics, design, synergic technologies, and implementation principles. Smart cities should satisfy minimal norms related to concepts such as smart governance, mobility, living, people, environment, and economy.

Smart cities are understood as enormously allocated CPS employing the help of sensors monitoring physical and cyber gauges and actuators that transfigure the environment in an effective way. Technology industries, governments, and organizations face the challenges resulting from aggressive urbanization. These bodies of authority and influence are creating solutions to improve urban life, relying on energy utilization or other services. Per the United Nations Population Prospects 2017 Revision report, the global urban population is exponentially expanding.

In 2018 54% of the global populace inhabited urban regions. Populations continue to increase in size and expand the spatial allotment of the global population. Thirty percent of the global population was in urban areas by 1950; the number is anticipated to be 66% by 2050 [24]. Taking into accounts the toll that civilization and an aging population levy on a city, urban regions have to reconsider their organizational structures and infrastructures accounting for new challenges that can be among the subjects of how to allocated resources efficiently, such as water, energy, raw materials, and food with the least waste possible. Fleet and spontaneous urban growth threaten civil development if devoid of proper infrastructure. Per this concept, the core areas of development for a smart city include lifestyle and quality of life, Information and Communication Technologies (ICT) communication, services and infrastructure, intelligence, people and society, environment and sustainability, governance and administration, economy and finance, and transportation and mobility.

Technically, the foundations of a smart city lie in the capacity of its digital governance, the efficiency of its infrastructure and water-smart grids, urban systems of mobility, buildings and structures, and data-related and public inclusion initiatives [25]. To summarize, an optimal smart city is a significant sociotechnical urban momentum of transformation revolutionizing how people live in and interact with the city, considering the fact that digital technology and data act as core enablers. Smart cities need to enhance how effectively they develop smart cities in the frames of large-scale CPSs as can be seen in cases such as Santander, Singapore, Dubai, Boston, and others [10].

4.3 Future trends in cyber-physical system within smart city

According to the industrial systems and different inception points, the urban population exceeded the rural population globally by 2018. Futuristic trends favoring urbanization are expected to increase aggressively within a decade; it is expected by 2050 that almost 70% of the global populace will be urban. Most cities will be housing over 10 million inhabitants and exhausting most of the global resources [18]. Smarter cities are necessary for coping with larger populations while remaining in shape socially and economically compared to their peers globally; these populations retain longer life spans and span more densely geographically [17].

This paper tackles the four significant aspects of a smart city in the future: smart grids, energy storage, building efficiency, and intelligent transportation systems. Firstly, a smart grid utilizes advanced cyberphysical software and technology in order to attain, evaluate, and perform actions based on information related to transmission, energy generation, allocation, and consumption. In doing so, smart grids improve the soundness, efficiency, and continuity of the network. Smart grids include smart meters, which are devices keeping track of energy consumption of buildings in real-time. The EU Member States require smart meters to reach 80% market penetration.

By 2030, 72% of European consumers will own smart electricity meters [18]. However, smart grids also rely on smart software to acquire, evaluate, illustrate, and then be in charge of appliances connected to the grid, masquerading as a virtual power plant. Smart grids give access of transmission and distribution to network operators so that they can perform accurate load forecasting and generate beneficial insights marking and adjusting possible errors within the grid and thus prioritize investment [32].

Secondly, energy storage includes mechanical, chemical, electrical, and thermal technological devices that provide electricity and energy for versatile implementations. These devices can be frequency and voltage control, peak shaving, and perpetuity of energy supply devices. Energy storage is a vital facilitator of a huge chunk of application of renewable energy within grids and allocated generation due to its frequency. Macrowise, around 10%–15% of losses occur during the allocation and transmission of electricity. Matching local generation and demand together via storage is quite constructive and productive; this can save up to 30%–40% of electricity consumption and hence bills.

Energy storage is vital to the smart city since it allows a local generation module that supplies direct feed to local buildings; thereby avoiding the hassle of passing through the national grid [7]. Thirdly, building efficiency focuses on the assimilation of a comprehensive set of cyberphysical software, technologies, and components within the frame of a physical environment to affect buildings' energy. The EU budgeted a target of a reduction of energy demand at 25% by 2030. Building efficiency does not just save costs; it enables officials to install monitoring, control technologies, switching energy consumption during times of peak demand [20].

Finally, intelligent transportation systems focus on software, technology, and physical infrastructure facilitating traveling around cities into a more efficient process, for example, electric vehicles, electronic payment systems, and new travel business models, such as car-sharing and carpooling. Intelligent transport systems pose a necessary answer to the ravaging economic impact of traffic congestion that is expected to penalize the UK ecò30, which is translated into a 63% increase of the gross cost in 2016. Multiple intelligent transport systems have been implemented across smart parking and cycling networks [26]. This section summarizes the main CPS directions in different domains through various studies and research efforts that have addressed the following:

1. Smart manufacturing: It is the application of infused hardware and software technologies to enhance the productivity of manufacturing products and delivery of services. It is considered one of the dominant CPS direction domains because of tweaks in domestic and international marketing, mass production, and economic boom. The characterizing of smart manufacturing was undertaken in the Industry 4.0 revolution that aimed to pioneer the manufacturing of the future [8].
2. Emergency response: It is responding to threats to public safety, health and welfare, and fending for the safety and integrity of natural assets, valuable infrastructures, and properties. CPS can devise and exercise a rapid emergency response through various sensor nodes in multiple areas, ready to respond to natural or man-made disasters. However, this rapid response requires the nodes to evaluate the situation in an aggregate and quick manner to keep the central authority in the loop effectively, even infrequently changing environments [27].
3. Air transportation: It is any military or civil aeronautics or aviation system along with their traffic management. Smart air vehicles, especially concerning military service, are expected to be dominant shortly. The crewless aerial vehicle, or "the drone," is a notable example of a smart air vehicle, an essential template for upcoming generations of aerial vehicles. CPS will create a lasting impact on the future of aviation and air traffic management [28].

4. Critical infrastructure: It is the set of valuable properties and public infrastructures necessary for the welfare and survival of the society, usually a standard among nations. The smart grid is a lucrative application in the vital infrastructure domain. It employs industrial and central power plants, renewable energy resources, energy storage and transmission facilities, energy allocation, and management facilities in smart homes and buildings [29].
5. Healthcare and medicine: These refer to the multiple issues concerning the physiological state of the patient. Specific attention is drawn to medical implementations in CPS research, opening venues for opportunities of research for the CPS community. These opportunities can be among the assisted living, technologies related to home care, smart medical devices, operating room technologies, and smart prescription [30].
6. Intelligent transportation: It is the confluence of advanced technologies of communication, sensing, computation, and control mechanisms in transportation systems to improve coordination, safety, and facilities in traffic management with real-time information sharing. These technologies expedite transportation, both ground and maritime, applying information sharing via satellites and master planning a communication environment among infrastructure, vehicles, and passengers' portable devices [28].
7. Robotic for service: It is the deployment of intelligent robots performing services for the well-being of humans in a fully autonomous, semiautonomous, or remotely controlled manner, excluding manufacturing operations. Robots can deploy, for example, for defense, environmental survey, control, logistics, and assist the living. Since next-generation robots will interact closely with humans physically, interpretation and learning of human activities by robots will become an important factor [31].
8. Building automation: It is the deployment of actuators, sensors, and control systems providing automation and optimum control over ventilation, heating, air conditioning, fire prevention, lighting, and security systems in buildings. Smart buildings will fulfill the model of smart grid and city concepts [20].

4.4 Research challenges and opportunities

The next generation of CPS smart cities research will provide a clear vision for the combination of the embedded systems with interconnected people and businesses through various innovative applications and services. The evolution of ingrained ICT systems in the smart city will be synergized together, creating highly distributed systems and offering innovations and opportunities in applications, technology, and business models [32].

In this section, CPS is not exhaustively covered but focuses on identifying essential challenges and opportunities for CPS in smart cities, providing complementary perspectives and focusing on CPS' impact on the critical areas of Big Data, cloud computing, and the Internet of Things (IoT).

4.4.1 Cyber-physical system in big data: challenges and opportunities

Big Data is a data set; both at static and dynamic states expanding over the boundaries of regular processing methodologies. It is highly diverse with multiple sources (variety), not one dimension and trusting and devoted to verification (veracity), enormous and sophisticated (volume), and being delivered rapidly and sporadically (velocity) [33].

Big Data extracts valuable information derived from data to use it intelligently. CPS acts as a master planner inceptor, collector, and allocator of data in a quick, verified, diverse, and immense manner. The meeting ground of Big Data and CPS expands, posing greater importance and impact thanks to the emphasis on data as irreplaceable business assets and crucial for businesses to remain competitive [11].

The challenges involve handling massive data production: new technologies will transform how massive input of data is handled and keep showing, in the case of CPS, what we know as a data tsunami. Besides this thriving volume, the diverse data from different sources will create the need for applications focusing on query, integration, analysis, high-performance computing devices, and methodologies for data reduction. This variety is creating different data stores to assist the changes from volatile data models [34].

Distributed data storage and processing: remote storage of Big Data ranks as one of the recurring concerns in terms of content and the technical field. Cloud-based models assist developers to reduce the costs of storing and processing Big Data compared to previous models, delivering data accessibility and IT empowerment, leading to decentralized data storage and processing some problems arising such as replication, parallelism, and requirements inherent to the natural attributes of CPS. Connected devices' number will expand rapidly and thus increase the sets of data and traffic. The necessity for real-time evaluation and accurate response will become increasingly crucial despite the increasing size of data and the faster velocity it occurs at [3].

Monetizing big data stemming from cps: the mix between Big Data and CPS is a bold move, from a management perspective. New approaches, though plenty, will be challenged by expanding market demands; it is necessary to optimize an enterprise's standing to be able to face these challenges. Healthcare or scientific research, domains that use massive datasets, have always been recurring users of Big Data. Combining Big Data with CPS creates lucrative new areas, for example, manufacturing or food production [10].

Data visualization: new means for data visualization are needed to deliver information for human decision-makers, due to the decentralized, volatile, and mobile nature of immense volumes of rapidly changing data sources. Standard visualization mediums, for example, control centers and dashboards, are not capable of keeping up with Big Data CPS setting [31].

While on the other hand, the opportunities include leveraging Big Data Analytics for CPS adaptation; CPS shall deliver vast volumes of data. Real-time data allows for unique opportunities for real-time planning and thus real-time decision-making; therefore this may be analyzed, driving the dynamic adaptation of CPS. It opens up room for techniques such as tweaking user-profiles and tracking and applying environmental context data through IoT. Big Data techniques will deliver useful insights on which related and appropriate elastic actions by CPS can be performed to rectify actions for those changes [22].

Greater customization and certification in products and services: customization in products and services are highly demanding. Big Data generates an opportunity to be filled with CPS and supposedly to create a beneficial economic effect. The embedded computing technology in different industries, for example, automotive, aerospace, healthcare, and transportation, gives a high starting point [35].

Assuring CPS assets stay online: enterprises need new impressive methods to stay competitive, effectively utilize their assets, and to assure that their assets remain online and account for unpredictable failures. Organizations should fully utilize the resources of available Big Data information, doing so through evaluating data and applying deduced insights to CPS. Proactive alerts on quality of products and concerns threatening reliability should eliminate interruption, unnecessary waste, and reputational damage. More intelligent and anticipative algorithms can enhance the capacity of tracking unscheduled maintenance, aging assets, saved money, reengineering processes, and boosted operational efficiency [11].

4.4.2 Cyber-physical system in cloud computing: challenges and opportunities

Cloud computing is a design that enables universal, comfortable, transparent, and on-demand admission to a joint resources pool of modifiable computing resources, for example, servers, applications, networks, services, and storage quickly delivered with scant managerial effort or service provider's interaction. For reducing costs, cloud computing helped alleviate capital expenditures and replace them with operational ones [25].

Cloud services now drive agility, productivity, and optimal performance, are a source of influence on the processes and many organizations, and a digitalization driver for mass users. Cloud computing delivers a convenient computing model for effortless integration of physical and computing components [22].

The challenges take in real-time data collection, analysis, and actuation: real-time collection of data is a necessary reality for CPS to excel. Subsequent analysis for and by CPS expedites appropriate decisions. Low-latency computation and actuation is a necessity within CPS due to their time-critical nature. CPS avert the need for analysis, evaluation, and decision-making to be in a central physical environment and do so through a well-allocated, robust cloud structure [18].

Multitenancy in CPS infrastructures: compatibility of systems to be placed in various places permits different clients to take part in the open-world setting of CPS. Integration of cloud resources as a component of high-performing and efficient CPS require special attention for servers and resources isolation reservation. It even surpasses server isolation and requires the proper adaptation of progressive cloud paradigms to correctly solve the isolation problem as a container or cloud-enabled and application–server isolation [8].

Dependable and predictable cloud Service-Level Agreements (SLAs) for CPS: it requires a high level of reliability, anticipation, planning, and tolerance to deploy critical CPS functionality within a cloud. Usually, cloud SLAs are provided based on best-effort only, making them barely provide a reliable mechanism for mitigation and risk prediction. An understanding is needed to comprehend the impacts of CPS workloads on infrastructure, cloud planning and risk management, audibility, and verifiable conformance to defined behavior regarding functional and nonfunctional requirements [36].

Cloud services and platforms for CPS construction and deployment: CPS relate to cloud infrastructures and critical administrative services. Therefore this type of cloud service is crucial for CPS construction and deployment; an essential feature of a cloud development framework for CPS is to back up the right development technique and methodology [37].

The opportunities cover scalability, elasticity, and availability: clouds can efficiently fine-tune the resources needed, varying with the loading volumes. CPS dictate resources needed and use devices that can be interconnected with the cloud entirely in a quick manner and hence the creation of a CPS ecosystem is relatively fast [28].

Additionally, the derivative quality of aggregated data, usually gauged by various CPS observers, could be considered to achieve more traceable and cost-effective cloud solutions. CPS acting as the basis for Cloud platforms, exercising replication and allocation of data on several sites, can exponentially spread globally. It enhances the availability and tolerance of CPS ecosystems, especially in emergencies [24].

Infrastructure costs reduction: maintaining cloud computing applications is cost-effective and convenient due to the lack of needing them to be available on each system; thus these applications are accessible from different locations. Multitenancy allows outsourcing of costs and resources across multiple users [38].

4.4.3 Cyber-physical system in Internet of Things: challenges and opportunities

New opportunities to efficiently manage sourcing, development, production, sales, and logistics through software-based services and improving upon novel business models for hybrid serviced products in the context of the IoT have been made available thanks to CPS. Beyond the physical flow of materials in fresh supply or end-product form, the information accumulated by implanted systems, smart items, sensors, and end-users orchestrate opportunities for constructing unique, crafted software-based services in IoT [39].

The value to the end-user manifests in software services and applications efficiently utilizing the information from sensors through systems-of-systems, whether in a direct or indirect manner. A more informed environment is available for decision-making since CPS sponsor admission to an exemplary amount of data about physical-world objects with low latency, improving organizational capacities [10].

The challenges in this context are close to instance-based architecture for a real business network of things: CPS dramatically allow simple handling of physical-world objects in software systems and services within IoT environments. For instance, rather than representing the physical items of a warehouse within a database through scanning or monitoring of physical items, the physical items themselves can be queried directly [28]. Such an instance-based architecture needs corresponding data standards that can be processed through an independent domain and specific services representing smart things, for example, products and batches, or designs [22], software-based services, business-wise, architect effortless access to the smart things [38], and deliver standardized services based on, triggering, or mirroring the movements between the participating organizations [35].

Additionally, this construct should be prompt to handle aggregations of trivial interactions and user authorizations, accessing and representing policies like a minimum stock level in logistics. Such a system based on instances could organically illustrate all relevant objects in a business network in a way to facilitate enterprises' agility in such volatile collaborations with various external parties and enable end-to-end visibility that traces and tracks the sequence of the smart things [1].

Service architecture for software-based services on top of CPS: futuristic services created from CPS will enable business users instead of guiding them, enhancing their approaches. Software-based services must consist of three attributes that differentiate them from current business applications: (1) to be self-explanatory so that business users are capable of designing them to be tailored to their needs without IT experts' support [17]; (2) to be flexible and adaptable so that they situate for all sizes of organizations [28]; (3) to be self-adapting, with flexible processes and direct integration with and of things into the network of business processes, in addition to being cross-organizational [23].

The process of creating associations of services or integrating extrinsic data is currently tiresome, time-consuming, and very costly, because users are not capable of implementing them without in-depth IT expertise. Creating a CPS architecture delivering services based on software and services in a simple manner, which enables business users to architect collaborations easily and quickly, is the real challenge to conquer nowadays [8].

Hypothetically, this process should be as simple as connecting to social networks, enabling end-users to custom-make their systems to address their needs or employ external or Big Data on the go, without IT help. A unified data model is required; it should allow users to focus on generic smart objects and services as per their individual needs within IoT space [33].

The opportunities encompass software-defined industries: the industry is appealing to just-in-time processing, production, and delivery of innovative products, goods, and services, often for a lot size of only one. Programmable facilities create tailored products, contrary to large quantities of products in the manufacturing process [8].

Highly customized and optimized manufacturing plants and supply systems will be capable of rectifying fluctuations and meeting customer demands effectively. In this model, the production relies on real-life demand gauging and reconfiguration of the methodologies of production in software via CPS instead of traditional long-term prediction and anticipation modules. Diversity in responding to demand becomes a critical point of strength in the market competition if an organization can take advantage of it [8].

Static lean manufacturing is going to be no more. Manufacturing enterprises should be more "Leagile" (lean + agile) or face extinction instead. "Leagile" evokes continuous monitoring and analysis of volumes of data related to production systems, inventories, and supply chains, while efficiently eliminating or minimizing waste [5]. Moreover, "Leagile" implies elastically shifting manufacturing potency on demand from the manufacturers at the edges;

those who wish to pool in their manufacturing prowess on the cloud to attract more orders from the ecosystem, increasing efficiency of their machines and resources material thus coping with the volatility of demand within the IoT world [18].

4.5 Visionary ideas for research trends

CPSs in the smart city context refer to different automation processes, for example, design, simulation, control, and verification. Upcoming CPS will connect, allocate, and coordinate in a manner that renders them responsive and robust. In a smart city, CPS can enhance productivity and quality via intelligent prognostics and diagnostics, thanks to Big Data from various machines, systems, and networked sensors [38].

Besides, CPS can fuse progress, achieved by extensive computing systems on modeling, planning, and forecasting using the surge of generated data during every day processes by many small data-driven devices, for example, sensors and actuators. Through contemporary technology trends including Big Data, cloud computing, and the IoT, CPS are becoming smoother, faster, and accelerated more than ever with significant impact on the smart city [30].

4.5.1 Cyber-physical system in big data: visionary ideas

Big Data for CPS design, evolution, and maintenance: many data sources generate a plethora of raw, unorganized data that can be evaluated, providing evidence on usage trends of CPS frameworks. In addition, data collected from CPS can be evaluated to track preferences, user trends, and features and performance improvement opportunities for advanced analytics, giving informed decision support for developing and evolving CPS. If those data sources were to be combined in meaningful ways, Big Data analytics can provide insights into the points in CPS and serve the information needs of other systems [17].

Big Data for CPS quality assurance and diagnosis: usage of automated analyses of CPS artifacts has been pondered on for some time. Today, owing to the expanse of data volumes and analytics capacities for enormous volumes of data, both structured and unstructured, CPS analytics harnesses new opportunities in the Big Data domain. Monitoring logs of a peculiar system of systems easily builds up to enormous data sizes in little time. Deviations and failure patterns can be analyzed through Big Data analytics, catering to such massive amounts of "metadata" being collected [12].

Big Data for CPS run-time monitoring and adaptation: Big Data employs the plethora of available data during the functioning of CPS monitoring of services, cloud infrastructure, things, and end-users shall provide information aplenty and in real-time promptly. It creates exclusive openings for real-time planning and decision-making, carving different paths for CPS adaptation. As an illustration, data techniques can deliver insights, founding a basis for actions that respond to changes based on changes in the CPS setting [18].

Optimizing CPS resource allocation: to optimize resources worldwide, installing shared data centers should be considered. Novel mechanisms are needed to adapt to handle such complicated global resource sharing. Efficient recycling of CPS units should be considered as a way to tweak resource usage and as a result, productivity [17].

Fostering CPS skill building during research activities: CPS research wherein industry and academia meet is to be deemed as an optimal, complementary method of delivering and educating functional people. CPS sponsor learning through various opportunities for designing work and joint research. Hence, CPS research should embolden their activities to promote these chances for skill building in CPS [10].

4.5.2 Cyber-physical system in cloud computing: visionary ideas

Dynamic adaptation of CPS: cloud computing facilitates the run-time adaptation of cloud infrastructures and applications to respond to context changes and system failures dynamically. Measurements of the execution parameters of infrastructure utilization and implementation, when collected in real-time and fused with the IoT, allows data execution imbued with insightful data concerning the system context delivered by numerous sensors. This run-time data provides extensive and more optimized opportunities to adjust the system automatically during operation continuously, for example, whether to scale the cloud horizontally or vertically [12].

The cloud as CPS: this means that it will be able to develop CPS running on a vast number of allocated resources as if they were running on a single resource while being able to tap into the nearly infinite power and being resilient to failures [11].

Leveraging CPS as key enabler: it is crucial to understand, practice, and authenticate assumptions and prototypes while providing analysis for related intellectual property. The viability discussion usually starts with understanding the value. Therefore the CPS design and the interplay are required and needed to combine with a robust, standardized, and feedback-driven approach. Design and the profession of design are crucial—the profession of design is currently redefined to create a desirable context [35].

Globally scaling CPS: for scaling initial success globally, new approaches driven from new mathematical models that benefit crowd designing are needed to make sure adoption is fostered. Complementarily, new concepts for catering to future needs for reactive design and cloud-based systems should manage continuous delivery and updates of content as well [39].

Pursue dynamic approach to CPS delivery: in an ideal scenario, technology resulting from CPS are preferred in packaged form, that is, in the form of application of frameworks or integrated models. CPS should bring their solution into a reusable frame for CPS perspective to foster adoption in practice [40].

4.5.3 Cyber-physical system in Internet of Things: visionary ideas

Perform CPS-IoT research exploiting real-world cases: addressing particular CPS-IoT problems allows the exploitation of outcomes. Real-world use has complex realistic scaling invoking problem understanding and appraisal. Rigorous and relevant empirical studies in the industry are necessities for the future of CPS. Real-world cases and data facilitate the understanding of applications of CPS practices. Addressing CPS-IoT challenges opens research difficulties to address and conquer the trench between theory and practice [35].

Deliver CPS pilots in IoT: this makes CPS results accessible as much as possible; CPS should construct and sustain preindustrial systems for its outcomes to attract industry premises. For sponsoring the understanding of these systems, existing technology can act as templates; innovative features should be presented once available in the CPS. These systems have to be interlinked to vital areas of CPS to open avenues for exploitation. Also, CPS design should be enhanced by virtual sessions that ease the understanding of how to work with the CPS and make it practicable; outcomes should be well recorded and result in justified standards for such records [33].

Larger-scale, integrated CPS is important for IoT research: the current work of CPS is extensive, integrated, and concentrates the efforts of research and industry domains, considering various scopes, angles, and aspects simultaneously to devise relevant and significant solutions in a practical manner. Constructing and authenticating new CPS methods to tend to the cognitive complexity, peculiar dynamicity, and scale of futuristic CPS needs larger-scale and longer-term joint efforts [15].

Pursuing an open IoT strategy for CPS research: this should be carried out early by tapping in to current ecosystems, sustaining and adopting open CPS outcomes; this develops an open-source community in the IoT. Ideally, CPS would become a part of an existing open source community to utilize the advantages of current systems. To ensure two-way communication, CPS has to remain compatible as much as possible, about current open-source applications and to integrate results into the open-source reference. Open source inspires development of open models and interfaces as a new, engaging direction. Open models serve as basic templates for standards allowing easy access for them later on. Open interfaces allow implementation by various vendors easily [13].

Modernizing CPS curricula: the need has been modernized to cater to the demands for basic and advanced CPS skills and prowess. Based on a sturdy foundation of CPS principles, curricula need to address contemporary technology trends, for example, data-intensive, cloud-based, and IoT- and CPS-oriented systems [33].

4.6 Roadmap of cyber-physical system strategic research

A research roadmap is essential to help the research community achieve its mission to create robust, sustainable, and quality-oriented CPS. CPS provide the solutions and services necessary to implement such scenarios of smart cities. CPS support a large-scale collection of operational data, through infrastructures and processes, ranging from the use of low-cost, across-the-board sensing technology to the support for secure uptake of information and feedback, whether direct or indirect, aggregated by citizens' data. To that end, the roadmap of strategic research for CPS of the smart city will, additionally, interoperate to integrate the smart services and help provide solutions for sustainable and efficient management of urban areas [32].

Intensify enabling sciences: CPS is built on scientific conclusions and technologies from different fields of sophisticated and large-scale technical and organizational systems. Therefore innovations in those fields are incentives for technology push factors [21].

Address human–machine interaction: CPS enable an acceptable interaction strictly with a wide range of human users. Therefore these systems have to understand the limits and capacities, assess the full reach and the expectations and intentions of their human users [17].

Sponsor cross-disciplinary research: to provide system-level services in CPS, features along the value-chain for these services should be masterfully integrated, whether concerning microsystem technology via software and systems engineering up to including macroeconomics [8].

Support maturation initiatives: the complexity of CPS necessitates the development of technologies that are mature in the context of forming large-scale CPS and demonstrate requirements of that relevant technology [5].

Promote the CPS infrastructure and ICT backbone that available at a large-scale with coordination of technical and organizational processes. Therefore performance information and communication infrastructure in the urban areas are core factors for their implementation [18].

Coordinate installation of key-systems: while CPS can address critical societal issues, their inherent attributes of large-scale systems surpassing national boundaries require a considerable investment in private and public infrastructure, for example, communicating stations, electric transmission, and distribution grid [34].

Provide reference platforms: defined levels in the stack of disciplines should create the means to influence innovation by building and facilitating the maturation, benchmarking, and support services integration in CPS [38].

Homogenize interoperability standards: CPS provide system-level functionality via the collaborative coordination of processes. The crucial regulatory and technical prerequisites for this association should be delivered without exhausting and limiting potential applications [9].

Define system-level design methodologies: added-value, innovative services provided by CPS usually stem from the cross-subsystem orchestration of more critical services implementing sequences of technical, physical, and organizational processes. Therefore broad-spectrum aspects should be considered when designing and constructing these services [8].

Provide open standards: CPS has to promote the definition, provision, and evolution of normative standards for coordination and support actions [33].

Promote open source and open license: as CPS sponsors the delivery of value-added services via the confluence of more basic services, access to interoperability platforms support is a fundamental prerequisite for rapid innovation and is entire of high importance for enterprises [22].

Increase open data: to optimize technical and social processes, CPS make full utilization of data for coordination and collaboration purposes. To quickly facilitate the implementation of innovative governing schemes, the availability of open data should be increased [16].

Harden infrastructure: CPS make use of open information and communication technology to coordinate the control of vital organizational and technical processes, including the electric grid (switches and power stations), systems and their roadside installation, marketplaces for energy trading, or traffic control centers [11].

Protect data ownership: using their deep embedding in sociotechnical environments, CPS sustain a substantial amount of delicate data, from profiling of traffic participants or patients to closely tracking sensitive production processes [1].

Adapt dependability regulations: CPS, owing to their characteristics, for example, cross-organizational, large-scale, and multiuser support, follow maintenance–update routines standardized in communication and information technology instead of the traditional design–implement standard cycle in the domain of dependable systems [1].

Stimulate collaboration: the innovative nature of CPS does not describe the technologies involved and the engineering challenge of systems of such a scale and diversity that demands advances in both theoretical foundation and practical engineering [15].

Provide awareness platforms: due to their infrastructure and set of vital services, futuristic CPS can create innovative added-value service-engineering, cross-discipline technologies to naturalize with practical and technical, rather than theoretical, capabilities [41].

Enable decision-making: decision-makers must understand the impact of CPS on society and business in order to make educated decisions; this sustains competitiveness and fosters the transfer of knowledge, which should be initiated [8].

4.7 The case study of a smart city

Urbanization retains many advantages enhancing economic and cultural activity and bringing ideas and people together. Closer people means less transportation lead times and higher productivity. Environmentalists postulate that efficiently planned cities benefit from less fossil fuel consumption and efficient allocation of land through expanding development density. Urbanization creates new human-made challenges. High populations exhaust the restricted spectra of resources of food, land, clean water, and energy. Though they promote economic productivity and a better quality of life, cities require systems focusing on efficiency and sustainability when it comes to resources [14].

Cities are faced with limited capital, and meager operating budgets must be utilized to cater to continually growing populations and increasing demand for services. The effect of urbanization is beyond calculation, evident in speculation that 66% of the global population will be urban by 2050. The ability to promptly manage to enlarge urbanization is vital. Cities must improve current and new services and infrastructure. For this to happen, it requires an abrupt change in infrastructure operations and management. The reach of intelligence and user-friendly interfaces of the various services provided by the city affect the lifestyle of residents. Technology contributes to this smartness [3].

Smart cities, as a concept, are still not a matter of agreement; there are roughly 120 different definitions of a smart city. The standard definition of a smart city, based on drawing from the power of data and the useful application of energy, technology, resources, and waste, is that it is a city that aspires to deliver a better version of life through engaging residents actively. Smart cities apply ICT to render standard infrastructures more durable, practical, inhabitable, and secure. The success of smart cities rely on the sharp implementation of all components of ICT, transparent, efficient, and central government that serve informed, engaged, and interconnected people [23]. Smart cities apply a four-step process:

1. Capture data: smart devices are responsible for collecting data in real time. For example, sensors along roads can aggregate information concerning road conditions and traffic jams while smart meters in offices and houses track consumption of electricity dynamically. Technological advances and the diminishing cost of devices render the wide spread of devices across cities feasible. These devices are the foundations of a smart city [11].
2. Communicate: collected data from smart devices and infrastructure is better communicated across control centers and servers. Smart cities need a communicative model facilitating interconnecting devices while ensuring scalability, integrity, interoperability, and privacy. Integrated communications strategies consist of service providers, communication infrastructure providers, IT vendors, and city governmental bodies [27].
3. Analyze: centralized data needs to be interpreted to deliver actionable insights. Data crunching necessitates computers and algorithms capable of processing and eligibly transform such data into intelligence. For instance, traffic sensors data show jams in certain areas and can offer alternative routes [3].
4. Act: as per this date, application of the analysis is needed for decisions. For example, a comprehensive electronic database of patient's medical records of patients captured by hospitals can influence strategic healthcare infrastructure decisions according to needs for certain medical services [13].

In less than five decades, cities have more than tripled in number from 548 back in 1970 to 1692 in 2015. With approximately four billion global people in cities today, a global wave of digital-era urbanization is occurring. Many urban agglomerates are doubling their smart city development efforts, affecting policy innovations and intensive global investments in new technology and data utilization approaches with the objective of sustained urban growth and social problems. The key pillars of a smart city are analyzing, in order to understand some of their chief features, tracking, and emphasizing good practices in each area as practical examples [3].

1. Telecom: telecom connectivity takes place centrally, forming the backbone of a smart city. Effortless and two-way connectivity is necessary for all components of smart cities. Cities have to utilize quick action and capacity and reliable networks. Smart cities require universal broadband connectivity through a fiber-optic, high-speed system and high bandwidth Wi-Fi networks that place consumer and smart cities within a speedy, efficient, and wholesome Internet. Naturally, these networks depend on machine-to-machine and machine-to-human communications. An estimated number of 50 billion devices will be online globally by 2020, requiring rapid and dependable infrastructures [4].

   To enable synergy of this vast number of devices connected to the Internet, international standards and reference architectures were shared and applied in a way that is needed to improve the communicative link between people and corporations within smart cities. Besides, privacy is a primary concern—as personal information is among the collected data. Cities need to execute privacy governance policies that define which data to acquire and store, the people with access to inspect which information from the data, and the extent of data rights and protections to which the parties involved are entitled. Another concern is security—considering the vast number of devices to be connected. Besides requirements for network firewalls, all end devices must be secured from attacks causing data theft or malfunctions [23].

2. Transportation: mobility and transportation challenge cities globally. Residents rely on transportation systems to commute, transport necessary goods, and travel. Cities deal with problems related to transportation daily. Expanding cities contend with congestion and slow trip times due to overburdened infrastructure, while older cities suffer from the adverse effects of deteriorating infrastructure. Transportation infrastructure needs years to build or rebuild [8].

Therefore smart traffic routing by smart sensors positioned along roads and traffic signals can track traffic flows. Back-end systems then analyze traffic streams and compute optimum interludes for traffic signals to ease congestion. Traffic signals can respond in real-time to enhance commutes. Smartphone apps can provide real-time traffic updates to passengers. Moreover, the wireless sensors ingrained in parking spots, detecting whether parking spaces are occupied, can be used for smart parking. This data can be transmitted to a central system and then sent to smartphones of people looking for parking spots. The massive volume of data collected via traffic and parking sensors, mobile phones, and smart cards can improve forecasting, predictive modeling, and simulations to devise plans for the infrastructure developments in the smart city [20].

3. Buildings: buildings are the siphons of energy and causatives of greenhouse gases in cities. According to the World Business Council for Sustainable Development, buildings caused 32% of global energy consumption and 19% of carbon emissions in 2018. As per this model, energy usage may double by 2050. Smart cities must consider new policies and technologies, rendering buildings efficient and minimally affect the environment, thus improving residents' quality of life and health [26]. Building information modeling (BIM) is remodeling the way buildings, utilities, and infrastructure are designed, planned, built, and managed. BIM is a model-based, intelligent methodology with insights enhancing planning, designing, constructing, and managing buildings and infrastructure innovatively.

Wireless sensors and meters in smart buildings aggregate information concerning aspects of the building, for example, lighting and light levels, energy consumption, ventilation, and humidity control. Implanted sensors within fan blades or lighting fixtures transmit data regularly to a central server and help enquiries for the needed data. This data, through the centralized building, is then used to influence the execution of sophisticated analytics. Modem BMS self-educate and anticipate predispositions for temperature, light, and a myriad of other services. Smart buildings apply integrated IT-enabled work order systems to facilitate work order flows and construct analysis based on parameters, that is, building, available and occupied areas, labor, personnel, materials, and different costs [19].

4. Utilities: water is one of the world's most valuable resources. Common issues can be, for example, diminishing water quality, shortages, and aging infrastructure. Cities have to contend with some of these water issues, if not all of them. Smart cities apply novel technologies to tweak water and wastewater management. Water sensors can smartly gauge pipe flow rates at different regions in the water pipe systems to pinpoint leakages places. Meanwhile, other sensors can determine water quality through specific parameters, for example, pH, oxidation–reduction potential, conductivity, dissolved oxygen, and turbidity in hard-to-reach areas [26].

These sensors can transfer data in real-time via standard or cellular networks to a central system. The central water management system can utilize sensors' data, detecting water pollution and leakages, planning priorities, and fixing schedules. The data can determine telltale issues with pipes, treatment plant issues, chemical spills, or the efficiency of water sanitization systems. Smart water meters can be available in all households; residents can monitor and compare their water consumption in real time in their neighborhood. Meters can transmit water usage information to water authorities, facilitating billing, and eliminate monthly manual meter readings [8].

On the other hand, energy is the city's paramount resource. Cities need to evolve in opposition to the growing demand for energy and render energy sources green and friendlier to the environment. At the center of smart energy initiatives implement smart grids and meters in cities. Smart electricity meters are sensor-based meters monitoring and tracking energy consumption in real time and giving consumers' feedback because of their usage and consumption. Smart meters installed in each office, home, and factory send such information in real time to smart grids. Smart grids aggregate and act on real-time information from energy suppliers and consumers, providing real-time reaction and monitoring, anticipation, and increased dependence.

Smart grids are self-repairing and can effectively isolate failing parts of the network, preventing blackouts and outages. Smart grids are easily applicable for integration within renewable sources, for example, wind farms, solar plants, and hydrostations, evolving the process of storage and distribution of energy throughout a smart city [22].

References

[1] D. Preuveneers, E. Ilie-Zudor, The intelligent industry of the future: a survey on emerging trends, research challenges and opportunities in Industry 4.0, J. Ambient. Intell. Smart Environ. 9 (3) (2017) 287–298.

[2] Y. Liao, F. Deschamps, R. Loures, P. Ramos, Past, present and future of Industry 4.0—a systematic literature review and research agenda proposal, Int. J. Prod. Res. 55 (12) (2017) 3609–3629.

[3] A. Ojo, Z. Dzhusupova, E. Curry, Exploring the nature of the smart cities research landscape, Smarter as the New Urban Agenda, Springer, Cham, 2016, pp. 23–47.

[4] H. Arasteh, V. Loia, A. Tomaseti, O. Troisi, M. Shafie, P. Siano, IoT-based smart cities: a survey, in: 16th International Conference on Environment & Electrical Engineering IEEE, pp. 1–6, 2016.

[5] A. Platzer, Logical Foundations of Cyber-Physical Systems, Springer, Heidelberg, 2018, pp. 1–639.

[6] K. Zhou, T. Liu, L. Zhou, Industry 4.0: towards future industrial opportunities and challenges, in: 12th International Conference on Fuzzy Systems Knowledge Discovery IEEE, pp. 2147–2152, 2015.

[7] R. Chaâri, F. Ellouze, A. Koubâa, B. Qureshi, N. Pereira, H. Youssef, et al., Cyber-physical systems clouds: a survey, Comput. Netw. 108 (2016) 260–278.

[8] L.D. Xu, L. Duan, Big data for cyber physical systems in industry 4.0: a survey, Enterp. Inf. Syst. 13 (2) (2019) 148–169.

[9] A. Seshia, S. Hu, W. Li, Q. Zhu, Design automation of cyber-physical systems: challenges, advances, and opportunities, IEEE Trans. Comput. Des. Integr. Circuits Syst. 36 (9) (2017) 1421–1434.

[10] U. Sivarajah, M. Kamal, Z. Irani, V. Weerakkody, Critical analysis of Big Data challenges and analytical methods, J. Bus. Res. 70 (2017) 263–286.

[11] D. Tokody, G. Schuster, Driving forces behind Smart city implementations—the next smart revolution, J. Emerg. Res. Solut. ICT 1 (2) (2016) 1–16.

[12] N. Silva, M. Khan, K. Han, Towards sustainable smart cities: a review of trends, architectures, components, open challenges in smart cities, Sustain. Cities Soc. 38 (2018) 697–713.

[13] Y. Ashibani, Q.H. Mahmoud, Cyber physical systems security: analysis, challenges and solutions, Comput. Security 68 (2017) 81–97.

[14] T.H. Kim, C. Ramos, S. Mohammed, Smart city and IoT, Future Gener Comp Sy. 76 (2017) 159–162.

[15] H. Han, S. Hawken, Introduction: innovation and identity in next-generation smart cities, City Cult. Soc. 12 (2018) 1–4.

[16] X. Jin, B.W. Wah, X. Cheng, Y. Wang, Significance and challenges of big data research, Big Data Res. 2 (2) (2015) 59–64.

[17] B. Varghese, R. Buyya, Next generation cloud computing: new trends and research directions, Future Gener. Comput. Syst. 79 (2018) 849–861.

[18] X. Wang, L.T. Yang, X. Xie, J. Jin, M.J. Deen, A cloud-edge computing framework for cyber-physical-social services, IEEE Commun. Mag. 55 (11) (2017) 80–85.

[19] M. Zhang, S. Ali, T. Yue, R. Norgren, O. Okariz, Uncertainty-wise cyber-physical system test modeling, Softw. Syst. Model. 18 (2) (2019) 1379–1418.

[20] S. Jeschke, C. Brecher, T. Meisen, D. Özdemir, T. Eschert, Industrial internet of things and cyber manufacturing systems, Industrial Internet of Things, Springer, Cham, 2017, pp. 3–19.

[21] Y. Liu, Y. Peng, B. Wang, S. Yao, Z. Liu, Review on cyber-physical systems, IEEE/CAA J. Automat. Sin. 4 (1) (2017) 27–40.

[22] A. Stankovic, Research directions for the internet of things, IEEE Internet Things J. 1 (1) (2014) 3–9.

[23] D. Serpanos, The cyber-physical systems revolution, Computer 51 (3) (2018) 70–73.

[24] V. Gunes, S. Peter, T. Givargis, F. Vahid, A survey on concepts, applications, and challenges in cyber-physical systems, KSII Trans. Internet Inf. Syst. 8 (12) (2014, 4242–4268.).

[25] D. Mourtzis, E. Vlachou, G. Dimitrakopoulos, V. Zogopoulos, Cyber-physical systems and education 4.0–the teaching factory 4.0 concept, Procedia Manuf. 23 (2018) 129–134.

[26] C. Yang, Q. Huang, Z. Li, K. Liu, F. Hu, Big Data and cloud computing: innovation opportunities and challenges, Int. J. Digital Earth 10 (1) (2017) 13–53.

[27] A. Gepp, K. Linenlucke, J. Neill, T. Smith, Big data techniques in auditing research and practice: current trends & future opportunities, J. Account. Lit. 40 (2018) 102–115.

[28] T.H.J. Uhlemann, C. Lehmann, R. Steinhilper, The digital twin: realizing the cyber-physical production system for industry 4.0, Procedia Cirp 61 (2017) 335–340.

[29] F. Zafar, A. Khan, R. Malik, F. Jamil, A survey of cloud computing data integrity schemes: design challenges, taxonomy and future trends, Comput. Security 65 (2017) 29–49.

[30] F. Ochoa, G. Fortino, G. Fatta, Cyber-physical systems, internet of things and big data, Future Gener Comp Sy. 75 (2017) 82–84.

[31] Y. Lu, Industry 4.0: a survey on technologies, applications and open research issues, J. Ind. Inf. Integr. 6 (2017) 1–10.

[32] R. Alguliyev, Y. Imamverdiyev, L. Sukhostat, Cyber-physical systems and their security issues, Comput. Ind. 100 (2018) 212–223.

[33] D. Xu, L. Xu, L. Li, Industry 4.0: state of the art and future trends, Int. J. Prod. Res. 56 (8) (2018) 2941–2962.

[34] R. Cowley, S. Joss, Y. Dayot, The smart city and its publics: insights from across six UK cities, J. Urban Res. Pract. 11 (1) (2018) 53–77.

[35] L. Rodríguez, L. Sánchez, L. García, G. Alor, A general perspective of Big Data: applications, tools, challenges and trends, J. Supercomput. 72 (8) (2016) 3073–3113.

[36] L. Mora, R. Bolici, M. Deakin, The first two decades of smart-city research: a bibliometric analysis, J. Urban Technol. 24 (1) (2017) 3–27.

[37] V.G. Kadam, S.C. Tamane, V.K. Solanki, Smart and connected cities through technologies, Big Data Analytics for Smart and Connected Cities, IGI Global, 2019, pp. 1–24.

[38] R. Petrolo, V. Loscri, N. Mitton, Towards a smart city based on cloud of things, a survey on the smart city vision and paradigms, Trans. Emerg. Telecommun. Technol. 28 (1) (2017) e2931.

[39] E. Ismagilova, L. Hughes, K. Dwivedi, Smart cities: advances in research an information systems perspective, Int. J. Inf. Manag. 47 (2019) 88–100.

[40] J. Moura, D. Hutchison, Review and analysis of networking challenges in cloud computing, J. Netw. Comput. Appl. 60 (2016) 113–129.

[41] D. Mishra, A. Gunasekaran, J. Childe, T. Papadopoulos, R. Dubey, S. Wamba, Vision, applications and future challenges of the Internet of Things: a bibliometric study of the recent literature, Ind. Manag. Data Syst. 116 (7) (2016) 1331–1355.

A new swarm intelligence framework for the Internet of Medical Things system in healthcare

Engy El-Shafeiy[1], Amr Abohany[2]

[1]COMPUTER SCIENCE DEPARTMENT, FACULTY OF COMPUTERS AND ARTIFICIAL INTELLIGENCE, SADAT CITY UNIVERSITY, SADAT CITY, EGYPT [2]INFORMATION SYSTEMS DEPARTMENT, FACULTY OF COMPUTERS AND INFORMATION SYSTEMS, KAFRELSHEIKH UNIVERSITY, KAFR EL SHEIKH, EGYPT

5.1 Introduction

Today, the Internet of Things is expanding rapidly as a part of big data. Millions of current hardware devices, such as wireless sensing sensor networks (WSNs) area units and smart devices [1,2], are predicted to be connected soon. WSNs can be used in multiple services and apps, most private and public institutions, particularly in the healthcare field and medicine. So, the data collected and gathered from the WSNs are regarded to be a major area of big data. Alternatively, multiple sensors around us produce sensor data that can be used, for example, intelligent transportation systems [3] are based on the assessment of a massive quantity of complicated sensor data. Enormous scope medicinal services applications [4] are particularly information costly as they contain a monstrous number of patients and administrations. One of the principle errands required to oversee huge detecting information streams is information mining. The information gathered from Internet of Medical Things (IoMT) is considered to have solid business sway. WSNs in healthcare insurance are the key empowering influences for the IoMT. Because of the quickening request on the IoMT objects in which the WSNs are the essential wellspring of enormous information, the greatly scaled information is gathered unreasonably, bringing about system traffic issues and consuming massive amounts of intensity and requiring huge measures of memory. This, in turn, will have an impact on the performance of the network. In addition, the great increase in the capabilities and numbers of sensors and smart sensor systems improves the use and generation of data. Current techniques for the identification and evaluation of medical conditions will have to be extended and revised. In healthcare, times are changing due to information technologies emerging. An increase in the numbers concerning the demographic changes of associate in nursing aging population in industrial nations is anticipated to put a major concern on

Swarm Intelligence for Resource Management in Internet of Things. DOI: https://doi.org/10.1016/B978-0-12-818287-1.00010-3

economies and healthcare systems [5,6]. Shortly, the retired people number (with an increased life expectancy [7]) will approach the number of working people worldwide [8] and the decreasing workforce will probably not be able to maintain the current support level to the elderly. In a healthcare insurance sense, there has been an expansion in the sum and styles of customer equipment and clinical sensors that gather some part of physical human wellbeing and mental highlight. The utilization of these advancements enables the end-clients (e.g., chronic patients) by offering the capacity to record and watch the status and, if the prerequisite emerges, get far off help. The quick development of detecting procedures and inserted figuring has opened up new open doors for unavoidable and unpretentious assortment of wellbeing information [9]. The wellbeing detecting advancements [10] for the most part guarantee to achieve the continuous chronicle and observing of the patients' human services status with high exactness, which gives incredible potential to lessen the bother and cost of patients' visits to the doctor. They are respected to be the foundation innovation for early ID and intercession of variations from the norm and productive patient administration [11]. In light of the highlights given, there are basically two principle sorts of sensors for the assortment of wellbeing information [12,13]: movement and physiological sensors. The previous are utilized to recognize and record physiological lists or states of being of patients for finding or treatment purposes, for instance, circulatory strain, blood glucose, electroencephalogram, electrocardiogram (ECG), electromy ography, and so on. An assortment of subtle wearable gadgets have been accounted for, for instance, remote neckband for ECG estimation [14], wearable circulatory strain meters and sleeves [15], cut free eyeglasses-based wearable gadgets for beat travel time and pulse observing [16], shirt based gadget for the estimation of blood vessel pulse [17], wearable photonic materials for beat oximetry [18], and a cell phone-based gadget for temperature estimation and wellbeing following rate [19]. These embodiments are capable of providing measurements ubiquitously and unobtrusively. Integration of various biomedical sensors can provide a variety of high-level health state monitoring functionalities. Examples of these are heart state sensing: heart disease is the primary cause of sudden death and thus sensing and tracking systems for heart diseases have attracted great attention. Heart signals include heart rate variability, heart rate, heart rate (RR), and P-QRS duration [20], which can be obtained from the patients' ECG [21]. An algorithm for monitoring the quality of ECGs was suggested [22], which proved that the quality of captured ECG recorded by mobile phone or wireless devices is good enough to diagnose heart disease. This research [23] suggested a personalized electrocardiogram (EKG) rhythm management system using a smartphone wireless medical sensor and the system can diagnose abnormal cardiac dysrhythmia. The system can diagnose and track abnormal cardiac dysrhythmia. Cheng and coworkers [24] reported real-time cardiovascular disease diagnosis on a smartphone, which identified the irregular rhythmic beating of the heart to indicate heart disease risks. Sleep sensing is not just a passive process, but also rather a highly dynamic process that is terminated by waking up, which highly represents the healthy state of people. Researchers performed sleep sensing using body sensor networks, wearable devices, and mobile phones. The IoMT-generated data analysis can allow the early detection of abnormal healthcare conditions (e.g., cognitively impaired people, elderly, patients). Moreover, the behavior profile parameters sensed by IoMT sensors provide knowledge for doctors to treat a specific patient. For example, identifying periodic changes in the body temperature or heart rate of a patient can be helpful data. Thus the finding of the frequent itemsets occurrence and the relationship between the physiological information obtained from the IoMT (association rules) can provide care to predict to the user. Association rules are a data mining technique that enables one to discover important knowledge from the extracted data that has also been utilized to analyze knowledge from the IoMT data [25]. Distributed and parallel big sensing data stream methods suppose that data are partitioned to the computing nodes in advance. Some methods have proposed distribution systems for big sensor data to process big sensing data stream by using Map-Reduce [26] to mine the search space in a distributed manner. It assumes a datacentric technique of distributed computing. Apache Hadoop is an open-source implementation of the Map-Reduce which needs the support of individual candidate node sets rather than the full dataset of the sensor. These multiple parallel mining techniques [26–28] have been suggested for the parallel construction of frequent itemsets trees and parallel mining of the tree structure in a distributed memory environment. Distributed mining of frequent itemsets using Map-Reduce shown in [28,29] are used in this chapter. The combination of wireless sensors and IoMT networks with mining and association rules research has developed a cross-disciplinary concept of ambient association rules to overcome the difficulties we face in everyday life [30]. This chapter proposes the use of IoMT data from WSNs in smart hospitals or homes to discover important activities of patients. Our study assumes that there are mechanisms for measuring useful data and using it in healthcare [31]. The proposed framework observes and analyzes readings from IoMT in smart hospitals to recognize activities and changes in behavior. In this chapter, the new framework, named swarm intelligence for the Internet of Medical Things (SIoMT), discovers useful data using these measures as an early stage to reduce the time taken. The chapter is organized as follows: Section 5.2 deals with the background for the IoMT in healthcare, explains knowledge discovery and data mining for the IoMT in healthcare, and vital measurements and parameters in healthcare. Section 5.3 explains swarm intelligence (SI) algorithms such as ant colony optimization (ACO). Section 5.4 deals with our framework, SIoMT, to analyze data in healthcare. Our framework is based on the adaptive ABC. The simulation results are given in Section 5.5 and the conclusion ends the chapter.

5.2 Background

5.2.1 The Internet of Medical Things

The IoMT is an Internet revolution in the field of medicinal services, where clinical gadgets/objects are masterminded normally, particularly for a huge scope to advance knowledge frameworks to empower setting related choices. The medical data is accessible by other things or components of complicated services. This IoMT system is integrated with computing and communications capabilities. The information rapid technology convergence occurs at technological innovation levels, namely, data, devices, and networks for communication.

The healthcare dependence on IoMT is aggregated to implement access to care, to enhance the care quality and to decrease the care cost.

IoMT innovation has been a typical technique for the usage and activity of human services applications lately. Since colossal information were created by huge amounts of appropriated clinical sensors, the securing, coordination, stockpiling, handling, and utilization of these information has become an earnest and essential issue for accomplishing objectives in social insurance. Referring to some related research [32–38], the characteristics of IoMT data in a wireless sensor network can be described as follows:

Multisource data: IoMT apps obtain data from multiple distributed medical sensors. These data types differ from integer to character, containing semistructured and unstructured data such as video streams, images, and audio. Mining these distributed data from multiple sources is essential for the growth of applications.

Huge scale dynamic medical: IoMT apps always link an enormous number of medical sensors. Communication between different objects in a large-scale dynamic environment produces a large volume of uninterrupted, high-speed, real-time, data streams. Scalable memory, filtering, and compression systems are therefore crucial for effective data processing.

Low-level with weak data: IoMT data from medical sensors are low-level with poor semantics before they are handled. The relationship between these medical sensors is a temporal–spatial connection. To implement these smart systems, the complex needs to be abstracted in an event-driven perspective from the mass of low-level data.

Inaccuracy medical data: Some studies indicate that most medical detecting devices can catch only semicorrect data due to an inaccurate reading, which creates problems for direct use. Multidimensional data analysis and handling are therefore essential for the broad implementation of IoMT apps.

Privacy and security sensitivity: IoMT data are typically associated with privacy sensitivities and stringent security requirements, particularly in the situation of IoMT applications involving the collection and processing of personal data and images [39].

5.2.2 Knowledge discovery and big data mining for the Internet of Medical Things in healthcare

The use of IoMT innovation in medication incorporates practically all angles, including clinical practice, far off observing, brilliant home consideration, clinical data, savvy emergency clinic emergency treatment, clinical hardware, clinical waste checking [40], blood donation center administration, and contamination control. A significant use of IoMT in clinical data innovation is clinical administrations, which depend on WSNs innovation and radio-recurrence identification.

Lately, the utilization of WSNs to distantly gather and screen the indispensable sign information of patients so as to gather information on their social insurance condition has become an effective arrangement in a shrewd emergency clinic to empower the developing number of old individuals and truly impeded individuals to get full help and care in the medical clinic.

5.2.3 Vital measurements and parameters in healthcare

According to a study, IoMT will become increasingly effective for medical care in the years ahead. Most doctors would use IoMT to store and exchange the data of patients. Above all, modern society's diseases, such as obesity, diabetes, asthma, or high blood pressure, are simpler to manage and treat in this manner.

Healthcare solutions could cushion the dramatic surge in healthcare costs by patients registering their vital signs themselves through IoMT at the hospital or home.

- Basal body temperature for body temperature. High temperature or fever can be a symptom of infectious disease, immunological disease, skin inflammation, tissue destruction, and cancers [41]. Hypothermia or low temperature can be a risk factor for chronic diseases, hypoglycemia, and trauma [42].
- Breathing rate. The normal breathing rate for an adult is 12–20 breaths/min. The value increases by the decreasing age. For a child, it is 15–30 breaths/min. And for infants and neonates, it is 25–50 breaths/min and 40–60 breaths/min, respectively. Breathing frequency is the number of breaths taken within a set amount of time (60 seconds) [43]. It helps to diagnose the abnormal state of the lung. Tachypnea can be caused by carbon monoxide poisoning, hemothorax, or pneumothorax [44,45].
- Blood pressure is the pressure exerted by circulating blood upon the walls of blood vessels [46]. High blood pressure or hypertension affects the kidneys, arteries, heart, or endocrine system. Low blood pressure or hypotension can lead to serious heart, endocrine, and neurological disorders. Severely low blood pressure can cause shock [47].
- The heartbeat is "the physical expansion of the artery" [48]. It is closely related to heart rate and cardiovascular disease. A small change of heartbeat indicates various medical conditions, for example, an infection or the evaluation of primary and secondary cardiomyopathy processes, among others.

5.2.3.1 Examples of chronic diseases

Some chronic diseases can gain more advantages from continuous IoMT (e.g., diabetes) than in others (e.g., cancer). The examples below are in the first category:

- Asthma. Asthma is a problem with the airways that carry air to our lungs. If the air tubes become narrower, it is difficult to breathe. In an asthma attack, we feel that we can't get enough air. Although asthma is common in children, most people with asthma are adults.
- Heart disease. The tubes that carry blood from and to our heart are called arteries and veins. As a consequence of bad feeding practices, lack of exercise, smoking, and family history, arteries can be partly closed. If a blood vessel gets too closed and the blood can't reach the other body parts, serious symptoms may appear and damage can occur. If the heart doesn't obtain sufficient blood as a consequence of a blockage, part of the heart muscle dies and the heart will no longer be able to pump. It is called a heart attack. If the brain doesn't obtain sufficient blood due to a blockage, a stroke may happen. A stroke

can kill us or damage our brain. Most people in Egypt die from heart disease than any other disease [49].

- Chronic disease mining. Chronic disease mining is a population-based and proactive approach that tackles chronic diseases early in the disease cycle to avoid disease development and minimize essential health complications [50]. Many methods can be used to enhance the health of all patients diagnosed under specific conditions, not just those visiting the provider's hospital. This strategy minimizes the subsequent need for acute interventions in the future and enables people to remain healthy and maintain their independence for as long as possible. Successful chronic disease mining framework is patient-centered, evidence-based, and involves various approaches and procedures. Chronic disease mining is a systematic approach to coordinating healthcare procedures through different levels.

The readings got by all IoMT (e.g., basal internal heat level (BBT), breathing rate (BR), circulatory strain (BP), and heart beat (HB)) can be appeared as a blend of the diverse biosignals from all IoMT (e.g., a biosignal understanding rundown). Contingent upon the reaches as well as sorts of boundaries, the qualities detected by each IoMT can be partitioned into a few classifications dependent on age and predefined run, as appeared in Table 5−1. For instance, if the readings from HB and BBT are HB-High and BBT-High at time Tn, individually, the biosignal perusing list for Tn would be as per the following: Tn (HB-high, BBT-high). In this manner, the readings consistently created by the IoMT for the patients can be appeared as:

- T1: BBT-basic high, BP-high;
- T2: BP-ordinary, HB-high, BR-exceptionally high;
- T3: BR-high, BBT-high;
- T4: HB-low, BBT-low, etc (Fig. 5−1).

5.3 Swarm intelligence

SI is the aggregate conduct of a self-organized and decentralized system that might be normal or artificial intelligence. This chapter briefly describes the two most broadly utilized SI calculations that are naturally inspired.

5.3.1 Ant colony optimization

ABC is a heuristic process within a relatively new field of research, SI [51]. This is a research direction exploring nature's behaviors and communication structures to successfully apply them to other practical and theoretical issues. Within the heuristics, SI joins the subsection of iterative improvement procedures. The subject of the investigation isn't only the behavior and the communication of the ants, but also the communication within bee-states [52,53].

The behavior of the members of a flock or a colony is based on the fact that they alone would not be able to survive or to achieve good solutions to occurring problems. Within these communities, there is a strict division of labor.

Table 5–1 Examples of biosignal values and their biomedical sensor abbreviations for adults and children.

		Adults range				Children range				Unit
A biosignal	**Abbreviations**	**Critical high**	**High**	**Normal**	**Low**	**Critical high**	**High**	**Normal**	**Low**	**Range unit**
Basal body temperature	BBT	Above 40°	39°–38°	37°–37.9°	36°–36.9°	Above 39°	38°–38.9	37°–37.5°	Below 37°	beats/min
Breathing rate	BR	21–25	17–20	12–16	Below 5	Above 25	21–25	12–20	Below 10	breaths/min
Blood pressure	BP	Above 110	90–109	65–84	35–59	Above 120	95–119	95–119	Below 40	mmHg
Heart beat	HB	Above 100	70–99	40–69	Below 40	Above 120	100–120	75–90	Below 50	beats/min

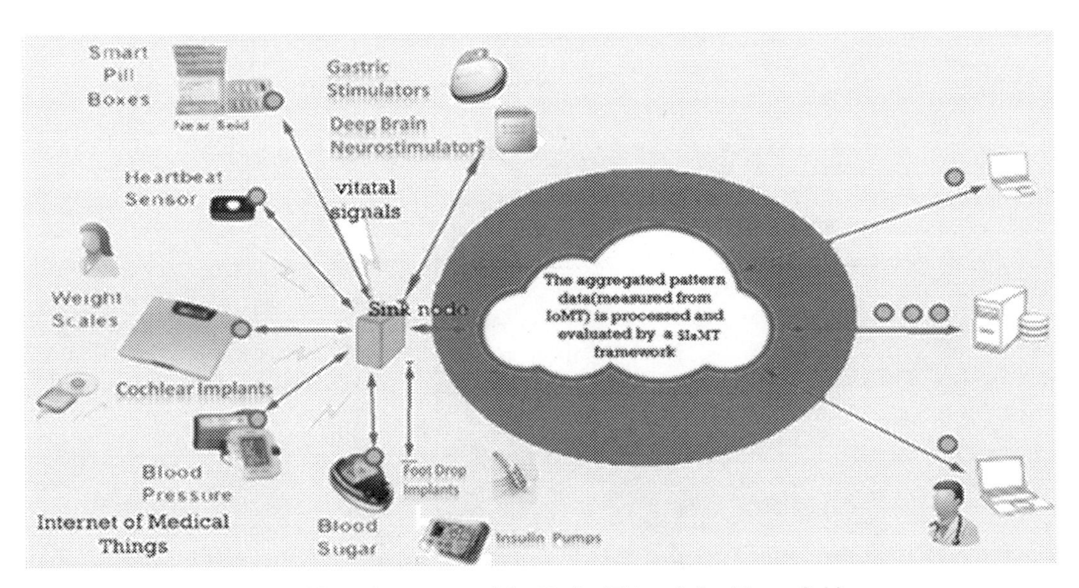

FIGURE 5–1 The data generated from the Internet of the Medical Things in healthcare fields.

The ABC, which is explained in more detail here, developed from the investigations and experiments of Goss et al. [54]. He developed the basic knowledge about behavior and cooperation within ant colonies. The starting point of his investigations was the behaviors of real ants in the wild. He observed that almost all the ants of a colony that are busy foraging use the same route to and from the feeding site. If it is blocked by an obstacle, for example, a rock, and it is impossible to continue to use this path, the ants can find a new shortest way to bypass this obstacle. This situation is shown in Fig. 5–2 (illustration based on [54]).

The observations in real ant colonies led to the well-known bridge experiment [54]. In the experimental setup, an ant colony was granted access to a food source. This can be achieved in two ways that differ only in the distance between the nest and the food source. The points B–H and H–D are each a unit length apart. The distance between B–C and C–D, however, is only half a length unit. The experimental setup is shown as an example in Fig. 5–3.

The key to explain this phenomenon lies in the indirect communication of the ants, the so-called stigmergy [54]. Any ant can mark a pathway with scents called pheromones. The attractiveness of a way is regulated by the quantity of a separate pheromone. Individual ants, moving almost randomly and encountering such a pheromone track, will track this lane. This ant can reinforce the existing trace by secreting pheromone. The amount of pheromone released may vary. This increases the attractiveness of a particular path with each ant that uses it. This results in positive feedback.

The amount of pheromone is shown in the right part of Fig. 5–3. At time $t = 0$, the search is still random, neither variant is labeled with an amount of pheromone. As the ants choosing the shorter path return to the nest, the pheromone value of the short variant increases

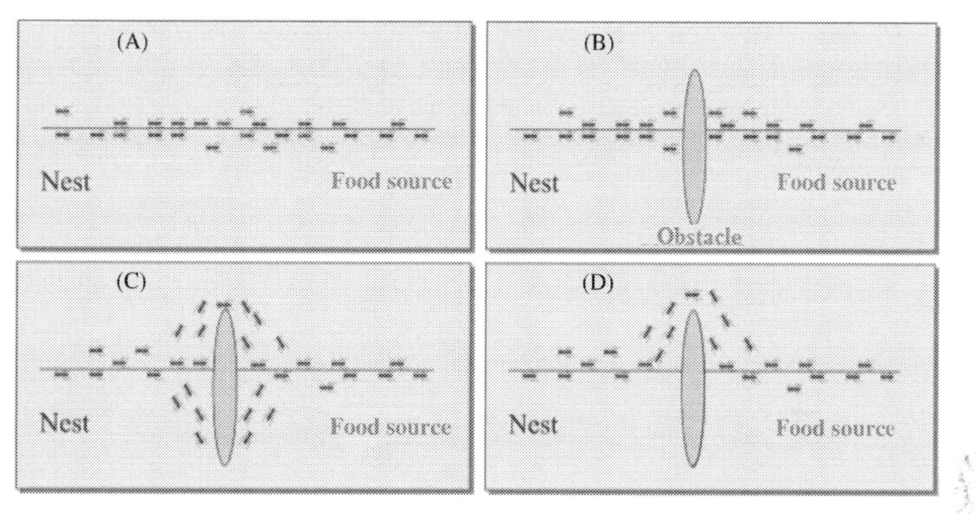

FIGURE 5–2 Foraging with obstacles (the ants can find a new shortest way to bypass this obstacle).

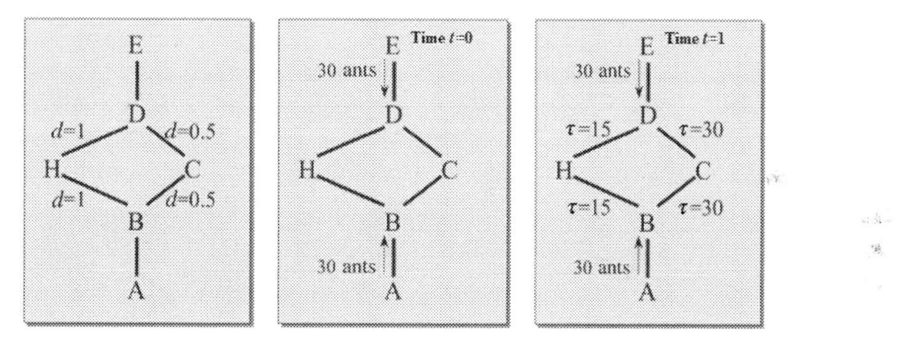

FIGURE 5–3 Bridge experiment.

faster and attractiveness increases. The shorter (30 units of pheromone) of the two variants is more attractive to an ant at time $t = 1$ than the longer (15 units of pheromone). A good solution to the problem can only be done through the entire colony, but not through a single ant.

5.3.2 Artificial bee colony algorithm

Implementation of the artificial bee colony algorithm has been used as inspiration to get a good overview of how the entire algorithm should be constructed as a pseudocode [55] and source code [56]. The algorithm for this work has then been modified to be adapted to road planning with obstacles. To create a complete ABC algorithm, random generation of roads has been implemented, which is discussed in detail in Section 2.2.1. The random generation receives a start node and destination node as input and generates a random path between these two nodes. A simple roulette wheel selection has also been implemented to represent the viewer selection, which bases the odds on the number of nodes a road contains. The fewer nodes the road has, the shorter the road, and the higher the chance that this road will be chosen by the spectator. This selection method has been chosen as it is the most commonly used method in the studies in Section 2.2.2. An important part of ABC is being able to generate neighboring solutions to the existing solutions. In a study using ABC for road planning [57], the authors have chosen to create neighboring roads by randomly selecting a section of the original road and then reusing the same feature that generated the original road. You let the start node by the first node in the selected section and the destination node is the last in the section. In this way, a new road is randomly dropped out within the locally selected section, making the neighboring road a similar route to the original one, albeit with a modification on the selected section. Fig. 5–4 shows what this might look like. A similar technique has been utilized in this work since it permits one to reuse a similar capacity as opposed to composing another capacity that generally meets a similar reason.

In outline, the calculation works so that roads are haphazardly produced and spared in a rundown. The working drones at that point create neighboring ways to each of these. A voracious choice is made between the neighboring street and the first road, where the road containing the least hubs, that is, the most limited course, is chosen. In the event that the neighboring road is the most brief course, it replaces the first road in the rundown. Watchers at that point select a course dependent on the roulette wheel determination and make a neighboring course to the chose course. Here as well, an insatiable determination is made. After each cycle, the most ideal way found so far is spared in a variable. In the event that a road in the rundown hasn't been supplanted by a neighboring road after the same number of cycles as the breaking point estimation of the working drones, this is supplanted by an investigation road that haphazardly creates a totally different road.

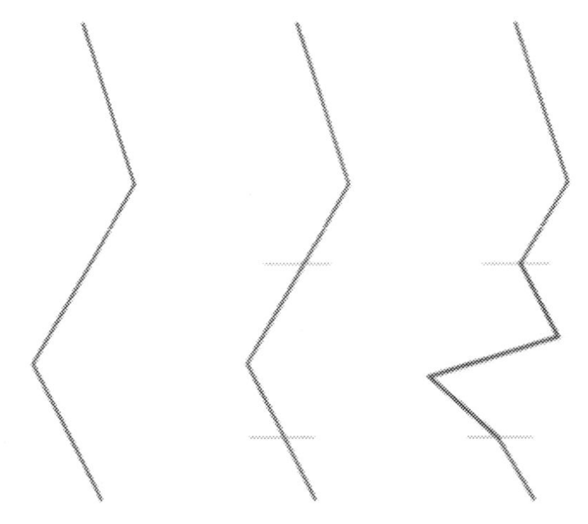

FIGURE 5–4 Demonstration of how neighboring streets are made. The left line is the first course. The two flat lines on the middle street are the haphazardly chosen area to be balanced and the red (dim dark in print form) line on the privilege is the arbitrary way between the chose segments. The correct street is a neighboring street produced from the left street.

5.4 The proposed named swarm intelligence for the Internet of Medical Things

A literature review was conducted on the topics related to this chapter. The main algorithms for grouping data related to the proposed algorithm were gathered and a comparative analysis of these algorithms was performed. According to this comparative study, it was noticed that the main problem detected in most of the presented algorithms is the fact that their results depend so much on important information that needs to be defined a priori (number of groups, the density of these groups, among others). As of centralizing structures, such structures are constantly accessed and modified throughout the grouping process and store information necessary for the correct formation of groups, making it impossible to apply these algorithms in inherently distributed domains, such as the IoMT where there are databases whose data are not centralized in a single one location for security reasons, privacy, among others.

In an attempt to overcome the abovementioned problems, this section describes in detail a SI inspired data grouping framework based on bee clustering for IoMT, which aims to form distributed groups of nodes with similar characteristics without any initial information related to the desired result, or even the need of the use of complex parameters. The framework (SIoMT) is inspired primarily by the automatic system of recruitment in the colony of the bees described. In nature, the bee performs dances as a goal to recruit other bees to the food that is a source of good quality nectar. This dance is used by bee clustering to form groups of nodes. However, in the proposed framework nodes dance to recruit new individuals to join their groups. Through this behavior, the framework SIoMT nodes can organize themselves into groups according to their characteristics or skills and in a distributed manner. The SIoMT framework is mainly inspired by the mathematical model of artificial bee colony recruitment described in Section 5.3.1. In nature, the bees perform the dance of recruitment to recruit other bees to collect food from a source of good quality nectar.

Thus each bee clustering node represents an object that needs to be grouped. The set of attributes of a given object constitutes the set of characteristics of this node that represents it. These characteristics are used in the group formation process to evaluate the similarity between the nodes. Fig. 5–5 shows an overview of how the whole grouping process happens. Each node has a set of possible states $\delta \in \mathcal{S} = \{d, v, w\}$ which indicate their actions as described below:

$\delta = d$: means the node is dancing to recruit/invite another node to be part of your group;

$\delta = v$: indicates that the node is moving another node;

$\delta = w$: means the node weight when watching the recruitment dance aiming to choose a new node to visit.

Node state change happens through some illustrated operation in Fig. 5–5. Table 5–2 summarizes the main parameters used by the SIoMT framework, where A, which represents the set of all nodes, has size $|A|$; X which represents the set of nodes characteristics is size $|X|$; $\mathcal{S}$, described above, represents the set of possible states of each node; P_a indicates the probability of one node abandon another according to their similarities, P_v indicates the

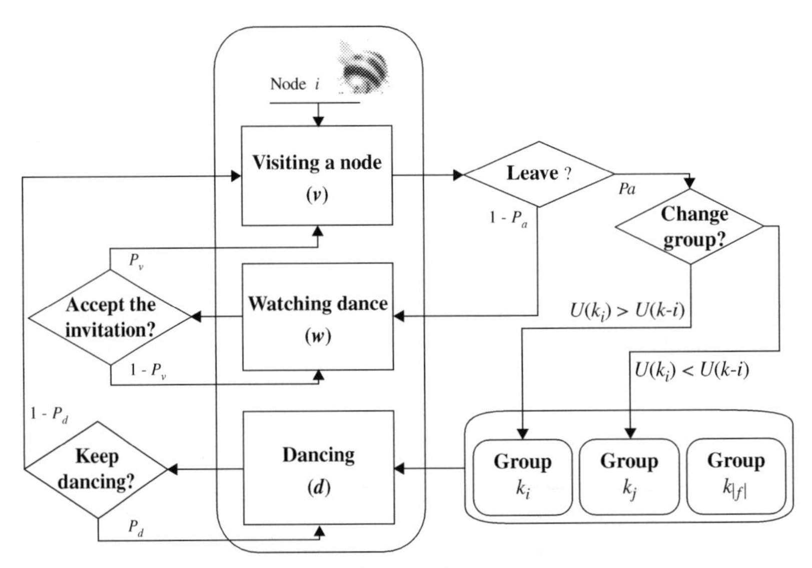

FIGURE 5–5 Clustering process performed by SIoMT framework.

Table 5–2 Main parameters of the SIoMT framework.

Parameter	Description
A	Node set (size $\|A\|$)
X	Node attribute set (size $\|X\|$)
$\mathcal{S} = \{v,w,d\}$	Node states v = visiting, w = node weight when watching, d = dancing
P_a	Probability of leaving node
P_v	Probability of visiting node P_d probability of continuing dancing to a group
$U(k)$	The utility of group k
N	Number of nodes in a group
D	Number of nodes dancing for a group
C	The centroid of a group
μ	Control the number of iterations of the algorithm

probability for one node to visit another, P_d represents the likelihood that the node will continue running the recruitment dance in order to invite other individuals to take part; $U(k)$ is the utility value of group k indicating the quality of this group; n indicates the number of nodes participating in a group; D represents the number of nodes who are performing the recruitment dance for a group in a particular instant t; c is a calculated element representing the centroid of a group; and μ is a parameter that controls the maximum number of iterations of the algorithm. The nodes of the proposed framework have limited knowledge regarding the other nodes of the model. Such knowledge is related to the set of their states

$\mathcal{S} = \{d, v, w\}$. When the state of the node is $\mathcal{S} = v$ or $\mathcal{S} = d$, it knows only those nodes that are allocated to your group at the current time (t), that is, once a given node no longer joins a group, it is forgotten by all the other elements of that group and becomes known only by participants in your new group. Similarly, a node who leaves a group does not store information related to its past for use later in its decision-making. When the node state is $\mathcal{S} = w$, it knows the individuals in your group and also about that subset of nodes whose state is $\mathcal{S} = d$.

At the beginning of the grouping process, each node is considered as a group. Next, visits are made between those who can check for similarities between them. During these visits, the nodes decide to stop joining their group to join the group of other nodes. In addition to deciding on group change, nodes must also decide on other actions that are important to the group formation process.

The main probabilistic choice the hub needs to make in the gathering procedure is whether to forsake the hub it is visiting. To make this decision, it is necessary for the node to check whether there are similarities between their characteristics and the characteristics of the node they are visiting. This check is performed using the Euclidean distance metric. Assuming then that node i is visiting node j, node i verifies how similar it is to node j by calculating the existing Euclidean distance in the joint attribute X attributes and the attributes of node j using Eq. (5–1), where $d(i, j)$ is the Euclidean distance between i and j and $|X|$ represents the number of attributes of i and j.

$$d(i,j) = \sqrt{\sum_{x}^{|x|} (i_x - j_x)^2} \qquad (5-1)$$

Hub i utilizes the outcome acquired by figuring the Euclidean separation, $d(i, j)$, as the likelihood of deserting hub j. For this, $d(i, j)$ is standardized and P_a is determined by Eq. (5–2), where $p|X|$ speaks to the biggest conceivable Euclidean separation between hubs, as their ascribes have been standardized to the stretch [0, 1].

$$P_a = \frac{d(i,j)}{\sqrt{|X|}} \qquad (5-2)$$

The lower the estimation of P_a, the more outlandish it is that hub i will surrender hub j, which implies that the two hubs have a high closeness list. Then again, the higher the incentive for P_a, the more prominent the probability of deserting j since as of now j are hubs with a low similitude list. It is important to note that once a decision on whether or not to abandon an individual is made, it is valid only for that instant t, that is, if in a future instant $t + 1$ the same individual is visited, the node checks again without any reference to the choices made in past moments. The decision that nodes must make about whether or not to abandon an individual is important because its outcome influences their next actions. It happens just before the node decides whether to switch groups to help them avoid wasting time deciding whether or not to join a group in which their individuals lack similar characteristics.

SIoMT system hubs need to choose whether or not to change to the hub bunch they are visiting. This dynamic happens at whatever point the probability of deserting P_a isn't met for a couple (i, j) of hubs. Hubs need to amplify the general value of grouping in an appropriated way. In SIoMT one attempts to boost a nearby utility identified with the gathering to which they have a place. In the proposed system one chooses whether or not to change bunches as indicated by the nature of their present gathering. The gathering nature of every hub is surveyed by an utility U that thinks about the qualities (traits) of all gathering members. The more comparable the individuals of a gathering, the bigger the utility of that gathering. Hence the utility worth $U(k_i)$ of gathering k_i, where hub i is found, must be augmented and determined by Eq. (5–3).

$$U(k_i) = 1 - var(k_i) \tag{5–3}$$

In Eq. (5–3), $var(k_i)$ indicates the intracluster variance of node i. However, for the calculation of U utility, only the variance of the group and not the variance of the resulting grouping are used. $var(k_i)$ needs to be minimized and can be calculated according to Eq. (5–4), where n is the number of elements within the group k_i, $d(i, c_i)$ is the Euclidean distance between node i and the centroid c_i of group k_i. The centroid c_i is calculated according to Eq. (5–5).

$$var(k_i) = \sqrt{\frac{1}{n} + \sum_{i \in k} d^2(i, ic)} \tag{5–4}$$

$$c_i = \frac{1}{n}\sum_{i \in c} i \tag{5–5}$$

To check whether to leave gathering k_i, hub i figures utility $U(k_i)$ also, utility $U - i$, which demonstrates the value of its gathering k_i without their support. In the event that $U(k_i)$ is more noteworthy than $U - i$, showing that the utility of gathering k_i is better with the nearness of i, hub i stays in gathering k_i. Something else, if $U(k_i)$ is under $U - i$, demonstrating that the value of the gathering of i is better without its support, i leaves k_i what's more, joins the gathering k_j of hub j that he is visiting. As a help, it is beyond the realm of imagination to expect to think about the nature of the gathering of the hubs they are visiting. Along these lines such data isn't utilized in the dynamic master cess about gathering change. For this choice, hub i just uses data about his own gathering, since it is expected that if the choice to leave j isn't fulfilled that Dad is most likely a hub like him, it might likewise be like hubs who have a place with hub j. In both of the past two circumstances, that is, regardless of whether to switch gatherings, i's next activity will be to play out the enrollment hit the dance floor with the objective of welcoming different hubs to join its gathering, regardless of whether this is another gathering or not. Toward the start of the move, hubs will probably need to choose whether or not to keep moving to enlist new people into their gatherings.

In SIoMT bunching, bunches are shaped because of a conduct displayed by hubs called enlistment move motivated by the scientific model portrayed. Similarly as in nature honey bees use move to welcome different mates to rummage on a decent quality nectar source, in the proposed system move is additionally utilized as a type of greeting to enroll others to join their gatherings. Along these lines at whatever point the hub's state is $\delta = d$ this shows the hub is moving to select others for his gathering. For this situation, he should choose whether or not to proceed in this state. How long a hub spends moving for his gathering k is a vital issue in the gathering procedure. In the event that hubs remain moving for an extensive stretch, the system may not work appropriately in light of the fact that all hubs will in general move while, is, they all will in general have a similar state $\delta = d$. Then again, if hubs remain moving for a brief period, the system meets to a bunch with an enormous number of little gatherings.

To control the time spent by the node dancing, a mechanism inspired by the division of labor model exhibited by social insects was used. As was described, nodes use a threshold-associated stimulus to probably decide whether to perform a task or not. In bees clustering nodes use this same stimulus S associated with a threshold θ to assist them in deciding whether or not to continue dancing for a particular group. The values of S and θ are used in the calculation of the probability P_d of continuing dancing according to Eq. (5–6).

$$P_d = \frac{s_i^2}{s_i^2 + \theta_i^2} \tag{5-6}$$

The update of the probability of continuing dancing P_d is directly related to the quality of the node group. Thus, the better the quality of node group i, the greater the value of P_d, that is, the greater the likelihood that i will continue dancing to recruit new nodes for his group. Group quality is measured by calculating the utility U of each group explained earlier. Since bunches are dynamic, that is, each cycle can get new hubs or likewise lose a few members, the likelihood P_d of hub i is refreshed by each cycle, while its state is $\delta = d$, as indicated by the accompanying principles. On the off chance that the utility $U(k_i)$ of gathering i of hub i at time t is more prominent than the utility $U(k_i)$ at time $t-1$, the boost of $i(S_i)$ is expanded and the edge of $i(\theta_i)$ is brought down by the steady α, consequently expanding the likelihood P_d of the hub to keep moving:

$$S_i = S_i + \alpha \tag{5-7}$$

$$\theta_i = \theta_i - \alpha \tag{5-8}$$

If the utility of the group $U(k_i)$ at time t is less than the utility $U(k_i)$ at time $t-1$, S_i is decreased and θ_i is increased by the constant α decreasing the probability P_d of the node to continue dancing Eqs. (5–7).

At whatever point P_d is fulfilled, the hub's state will remain with the goal that he can keep moving and enlisting new people to his gathering. At the point when P_d isn't fulfilled, showing that the gathering's engaging quality is diminishing, the hub finishes up the enlistment move however stays in a similar gathering. While the system is moving on, those nodes that are assisting their dance are intended to be guided to a group that contains individuals more similar to them than members of their current group. Such nodes who are watching the recruitment dance have state $\delta = w$ and also need to make a decision.

$$P_v = \frac{D(K_I)}{n} \quad (5-9)$$

As shown in Fig. 5–6 another moment of decision happens when node i is in the state $\delta = w$, which means that he is watching the recruitment dance performed by other nodes of the colony. At this moment, i randomly draws, with uniformly distributed probability, node j whose state is $\delta = d$, and visits this node j with probability P_v. To decide probabilistically whether or not to visit node j, i calculates the probability P_v according to Eq. (5–9), where

```
Framework SIoMT
initialize the colony with |𝒜| bees;
initialize parameter α;
repeat for each step
    foreach i ∈ |𝒜| do
        if step = 0 then
            randomly choose Node j;
            δi ← v;
        if δi = v then
            calculate Pa (Equation 2);
            if Pa then
                δi ← w;
            else
                calculate utility U(ki) and U − i (Equation 3);
                if U(ki) < U(k − i) then
                    kj ← i;
                δi ← d;
        else if δi = d then
            calculate Pd (Equation 6);
            if Pd then
                keeping dancing;
            else
                δi ← v;
        else if δi = w then
            repeat randomly choose a bee dancer j with state δj = d
                calculate Pv (Equation 7);
            until Pv;
            δi ← v;
until maxSteps;
```

FIGURE 5–6 The flowchart to the SIoMT framework.

$D(k_j)$ is the number of nodes dancing to recruit nodes for group k_j and n is the number of nodes that belong to node group j.

Thus the greater the number of nodes dancing for a given group, the greater the chance that other nodes will test this group.

The framework starts with the creation of the colony and its nodes and the initialization of parameter α, o which is used to update the stimulus S and the q threshold θ of each node in the calculation of probability P_d.

After the startup phase, nodes begin the group formation process. Visits between nodes are provided by invitations made by nodes whose status is $\S = d$. However, in the initial step (line 5) when no node is dancing to make the invitations, it is necessary for each node to randomly draw the first node to visit. All nodes are likely $\frac{1}{|X|}$ to be drawn.

This draw happens only in step 0 of the grouping process, in the others the visits happen because some nodes will perform the recruitment dance. At this time 0, the state of all nodes is $\S = v$ (line 7).

On the off chance that the condition of hub i is $\S = w$ (line 23), it implies that i is watching the move of enlistment, that is, i is watching those hubs whose state is $\S = d$. So he will arbitrarily draw another hub j to visit. All hubs with state $d = d$ are similarly liable to be drawn by i. Subsequent to pooling hub j, i should choose whether or not to acknowledge this present hub's greeting. To make this decision, i calculates a probability P_v according to Eq. (5–7) which considers the proportion of nodes that are dancing for the j group in relation to all nodes who are participating in that group. If i visits j, the state of node i becomes $\S = v$ indicating that in the next step the grouping process restarts and i's action will be to visit the node already raffled off and check whether to leave it or not. If, to the contrary, node i doesn't visit j, i will continue to raffle other nodes to visit. The bee clustering framework will execute until the maximum number of iterations (max steps) are achieved.

5.5 Performance evaluation of swarm intelligence for the Internet of Medical Things framework

It is important to note that although medical sensor readings were performed with nodes, the final version of the code uses a random number library to generate shared data on the network. Although some tests were performed using the medical sensors purchased for this purpose, a random number library providing floating-point values as well as the most common medical sensors. The use of this library is justified due to the fact that changes in basal body temperature, breathing rate, blood pressure, and heartbeat require manipulation of the internal environment, something extremely complex to simulate in the laboratory.

In order to verify the performance of the proposed framework in performing the grouping of the IoMT, some public were used, to build a medical-care system by the wireless sensor network. A sensor network of nodes was applied to process the vital signals and grouped the results through wireless devices.

To check the nature of the proposed structure, a few examinations were performed with the information depicted. The evaluation of the grouping obtained with the use of bee clustering was performed by calculating the F-measure and Rand index.

The values used for the parameters were: $\alpha = 0{:}02$, max steps $= |A| \times \mu$ and $\mu = 6$ (Table 5–3).

These qualities were picked after a few tests with various qualities for every boundary. The quantity of hubs previously utilized and the quantity of qualities $|X|$ for every hub separately get the base size qualities information and the quantity of characteristics of every information. Along these lines, $|A|$ and $|X|$ have various qualities for each base to be utilized.

Fig. 5–7 shows the diagram with the quantity of hubs in each state during the running of a reenactment. As can be seen, the state $\mathcal{S} = d$ is what focuses a more prominent number of hubs all through the recreation. This is on the grounds that hubs are taking an interest in great quality gatherings which make them move to enroll different hubs to be a piece of their gatherings. In the first steps of the simulation, the number of nodes in state $\mathcal{S} = w$ exceeds the number of nodes in state $\mathcal{S} = v$ because the number of visits that result in abandonment are large, which makes nodes have to attend the recruitment dance to pick new individuals to visit.

Table 5–3 Average F-measure resulting from 30 repetitions for the honey bee algorithm, swarm clustering, and SIoMT framework.

	Honey bee algorithm	Swarm clustering	SIoMT framework
F-measure	0.8915	0.923	0.985
Rand measure	0.8798	0.924	0.978
Number of groups	9	10	10.4
Identified	9	10	10.4

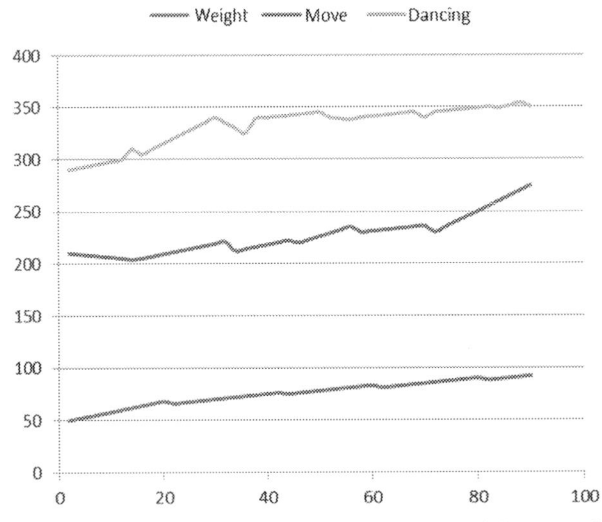

FIGURE 5–7 Number of nodes in each state $\mathcal{S} = \{w; v; d\}$ during a simulation of the SIoMT framework.

5.6 Conclusions

This chapter presented an efficient framework for the analysis of the periodic data cluster obtained from IoMT. A correlation test was implemented on the acquired cluster nodes and improved treatment and smart healthcare, especially for seniors in the smart hospital. The developed framework, named SIoMT, can discover and manage useful-data grouped using these measures at an early stage by developing the bee colony optimization, and this reduced the time. The use of the SIoMT framework for the distributed nodes has been widely explored. So, SIoMT became the de facto framework for the analysis and management of the data group. Furthermore, some experiments have been performed to validate the effectiveness of the proposed SIoMT framework.

References

[1] T. Dhope, D. Simunic, M. Djurek, Application of DOA estimation algorithms in smart antenna systems, Stud. Inform. Control. 19 (4) (2010) 445–452.

[2] D. Merezeanu, G. Vasilescu, R. Dobrescu, Context-aware control platform for sensor network integration in IoT and Cloud, Stud. Inform. Control. 25 (4) (2016) 489–498.

[3] J. Zhang, F.Y. Wang, K. Wang, W.H. Lin, X. Xu, C. Chen, Data-driven intelligent transportation systems: a survey, IEEE Trans. Intell. Transport. Syst. 12 (4) (2011) 1624–1639.

[4] Bekiri, Roumaissa, Abdelhamid Djeffal, and Messaoud Hettiri. A Remote Medical Monitoring System Based on Data Mining. No. 3359. EasyChair, 2020.

[5] L.M. Anderson, S.C. Scrimshaw, M.T. Fullilove, J.E. Fielding, J. Normand, Culturally competent healthcare systems: a systematic review, Am. J. Prevent. Med. 24 (3) (2003) 68–79.

[6] J. Macinko, B. Starfield, L. Shi, The contribution of primary care systems to health outcomes within Organization for Economic Cooperation and Development (OECD) countries, 1970–1998, Health Serv. Res. 38 (3) (2003) 831–865.

[7] B.L. Neugarten, The future and the young-old, Gerontologist 15 (1_Part_2) (1975) 4–9.

[8] D.W. DeLong, Lost Knowledge: Confronting the Threat of an Aging Workforce, Oxford University Press, 2004.

[9] Tam, Nguyen Thi, et al., Multifactorial evolutionary optimization for maximizing data aggregation tree lifetime in wireless sensor networks. Artificial Intelligence and Machine Learning for Multi-Domain Operations Applications II. Vol. 11413. International Society for Optics and Photonics, 2020.

[10] Mahajan, Sachit, et al., A citizen science approach for enhancing public understanding of air pollution. Sustainable Cities and Society 52 (2020): 101800.

[11] Breteler, Martine JM, et al., Are current wireless monitoring systems capable of detecting adverse events in high-risk surgical patients? A descriptive study. Injury 51 (2020): S97–S105.

[12] Pires, Ivan Miguel, et al., A Research on the Classification and Applicability of the Mobile Health Applications. Journal of Personalized Medicine 10.1 (2020): 11.

[13] Wang, Weichen, et al., Social Sensing: Assessing Social Functioning of Patients Living with Schizophrenia using Mobile Phone Sensing. Proceedings of the 2020 CHI Conference on Human Factors in Computing Systems. 2020.

[14] Ekerete, Idongesit, et al., Unobtrusive Sensing Solution for Post-stroke Rehabilitation. Smart Assisted Living. Springer, Cham, 2020. 43–62.

[15] Lou, Zheng, et al. Reviews of wearable healthcare systems: Materials, devices and system integration. Materials Science and Engineering: R: Reports 140 (2020): 100523.

[16] Lee, Gaang, et al., Wearable Biosensor and Hotspot Analysis–Based Framework to Detect Stress Hotspots for Advancing Elderly's Mobility. Journal of Management in Engineering 36.3 (2020): 04020010.

[17] Talukder, Md Shamim, et al., Predicting antecedents of wearable healthcare technology acceptance by elderly: A combined SEM-Neural Network approach. Technological Forecasting and Social Change 150 (2020): 119793.

[18] Ray, Lydia. Cyber-Physical Systems: An Overview of Design Process, Applications, and Security. Cyber Warfare and Terrorism: Concepts, Methodologies, Tools, and Applications. IGI Global, 2020. 128–150.

[19] Tuli, Shreshth, et al., Next generation technologies for smart healthcare: Challenges, vision, model, trends and future directions. Internet Technology Letters 3.2 (2020): e145.

[20] D. Bansal, M. Khan, A.K. Salhan, A review of measurement and analysis of heart rate variability, in: Computer and Automation Engineering, 2009. ICCAE'09. International Conference on, pp. 243–246. IEEE, 2009.

[21] T. Ince, S. Kiranyaz, M. Gabbouj, A generic and robust system for automated patient-specific classification of ECG signals, IEEE Trans. Biomed. Eng. 56 (5) (2009) 1415–1426.

[22] M.A. Kabir, C. Shahnaz, Denoising of ECG signals based on noise reduction algorithms in EMD and wavelet domains, Biomed. Signal. Process. Control. 7 (5) (2012) 481–489.

[23] V. Gay, P. Leijdekkers, A health monitoring system using smart phones and wearable sensors, Int. J. ARM 8 (2) (2007) 29–35.

[24] J.J. Oresko, Z. Jin, J. Cheng, S. Huang, Y. Sun, H. Duschl, et al., A wearable smartphone-based platform for real-time cardiovascular disease detection via electrocardiogram processing, IEEE Trans. Inf. Technol. Biomed. 14 (3) (2010) 734–740.

[25] G. Hripcsak, S. Bakken, P.D. Stetson, V.L. Patel, Mining complex clinical data for patient safety research: a framework for event discovery, J. Biomed. Inform. 36 (1–2) (2003) 120–130.

[26] L. Zhou, Z. Zhong, J. Chang, J. Li, J.Z. Huang, S. Feng, Balanced parallel fp-growth with mapreduce, in: Information Computing and Telecommunications (YC-ICT), 2010 IEEE Youth Conference on, pp. 243–246, IEEE, 2010.

[27] K.L. Tan, Z.H. Sun, An algorithm for mining FP-trees in parallel, Comput. Eng. Apps 13 (2006) 155–157.

[28] Y.K. Woon, W.K. Ng, E.P. Lim, A support-ordered trie for fast frequent itemset discovery, IEEE Trans. Knowl. Data Eng. 16 (7) (2004) 875–879.

[29] E.A. El-Shafeiy, A.I. El-Desouky, A big data framework for mining sensor data using hadoop, Stud. Inform. Control. 26 (3) (2017) 365–376.

[30] R. Srikant, R. Agrawal, Mining generalized association rules, 1995.

[31] D.L. Streiner, G.R. Norman, J. Cairney, Health Measurement Scales: A Practical Guide to Their Development and Use, Oxford University Press, USA, 2015.

[32] E.L. Engy, E.L. Ali, E.G. Sally, An optimized artificial neural network approach based on sperm whale optimization algorithm for predicting fertility quality, Stu.d Inf. Control 27 (3) (2018) 349–358.

[33] E.A. El-Shafeiy, A.I. El-Desouky, S.M. Elghamrawy, Prediction of liver diseases based on machine learning technique for big data, International Conference on Advanced Machine Learning Technologies and Applications, Springer, Cham, 2018, pp. 362–374.

[34] W.B.A. Karaa (Ed.), Biomedical Image Analysis and Mining Techniques for Improved Health Outcomes, IGI Global, 2015.

[35] N. Dey, A. Ashour, A.F. Shi, V.E. Balas, Soft Computing Based Medical Image Analysis, N.p.: 16 January 2018. Print.

[36] S.A. Fadlallah, A.S. Ashour, N. Dey, Advanced titanium surfaces and its alloys for orthopedic and dental apps based on digital SEM imaging analysis, Adv. Surf. Eng. Mater. (2016) 517–560.

[37] L. Moraru, S. Moldovanu, A.L. Culea?Florescu, D. Bibicu, A.S. Ashour, N. Dey, Texture analysis of parasitological liver fibrosis images, Microsc. Res. Technol. 80 (8) (2017) 862–869.

[38] S. Hore, S. Chakroborty, A.S. Ashour, N. Dey, A.S. Ashour, D. Sifaki-Pistolla, et al., Finding contours of hippocampus brain cell using microscopic image analysis, J. Adv. Microsc. Res. 10 (2) (2015) 93–103.

[39] D. Nandi, A.S. Ashour, S. Samanta, S. Chakraborty, M.A. Salem, N. Dey, Principal component analysis in medical image processing: a study, Int. J. Image Min. 1 (1) (2015) 65–86.

[40] S. Kamal, N. Dey, A.S. Ashour, S. Ripon, V.E. Balas, M.S. Kaysar, FbMapping: an automated system for monitoring Facebook data, Neural Netw. World 27 (1) (2017) 27.

[41] E.H. Page, N.H. Shear, Temperature-dependent skin disorders, J. Am. Acad. Dermatol. 18 (5) (1988) 1003–1019.

[42] D.F. Danzl, R.S. Pozos, Accidental hypothermia, N. Engl. J. Med. 331 (26) (1994) 1756–1760.

[43] B.M. Saykrs, Analysis of heart rate variability, Ergonomics 16 (1) (1973) 17–32.

[44] S. Yurtseven, A. Arslan, U. Eryigit, M. Gunaydin, O. Tatli, F. Ozsahin, et al., Analysis of patients presenting to the emergency department with carbon monoxide intoxication, Turkish J. Emerg. Med. 15 (4) (2015) 159–162.

[45] N.M. Hawkins, M.C. Petrie, P.S. Jhund, G.W. Chalmers, F.G. Dunn, J.J. McMurray, Heart failure and chronic obstructive pulmonary disease: diagnostic pitfalls and epidemiology, Eur. J. Heart Fail. 11 (2) (2009) 130–139.

[46] A.C. Burton, Relation of structure to function of the tissues of the wall of blood vessels, Physiol. Rev. 34 (4) (1954) 619–642.

[47] A. Blalock, Experimental shock: the cause of the low blood pressure produced by muscle injury, Arch. Surg. 20 (6) (1930) 959–996.

[48] R.M. Ferrari, C.S. Taylor, J.W. Lasersohn, F.J. Benetti, J.J. Akin, R. Ginn, et al., U.S. Patent No. 5,875,782. Washington, DC: U.S. Patent and Trademark Office, 1999.

[49] J. Stamler, Epidemiology of coronary heart disease, Med. Clin. North. Am. 57 (1) (1973) 5–46.

[50] J.B. Soriano, J. Zielinski, D. Price, Screening for and early detection of chronic obstructive pulmonary disease, Lancet 374 (9691) (2009) 721–732.

[51] E. Bonabeau, D.D.R.D.F. Marco, M. Dorigo, G. Theraulaz, Swarm Intelligence: From Natural to Artificial Systems, Oxford University Press, 1999. No. 1.

[52] S. Camazine, J.L. Deneubourg, N.R. Franks, J. Sneyd, E. Bonabeau, G. Theraula, Self-Organization in Biological Systems, Princeton University Press, 2003.

[53] T.D. Seeley, S. Camazine, J. Sneyd, Collective decision-making in honey bees: how colonies choose among nectar sources, Behav. Ecol. Sociobiol. 28 (4) (1991) 277–290.

[54] S. Goss, S. Aron, J.L. Deneubourg, J.M. Pasteels, Self-organized shortcuts in the Argentine ant, Naturwissenschaften 76 (12) (1989) 579–581.

[55] J. Lee, Vägplanering i dataspel med hjälp av Artificial Bee Colony Algorithm, 2015.

[56] D. Karaboga, B. Basturk, A powerful and efficient algorithm for numerical function optimization: artificial bee colony (ABC) algorithm, J. Glob. Optim. 39 (3) (2007) 459–471.

[57] K.A. Eldrandaly, M.M. Hassan, N.M. AbdelAziz, A modified artificial bee colony algorithm for solving least-cost path problem in raster GIS, Appl. Math. Inf. Sci. 9 (1) (2015) 147.

6

Current vulnerabilities, challenges and attacks on routing protocols for mobile ad hoc network: a review

Mazoon AlRubaiei[1], Hothefa sh Jassim[1], Baraa T. Sharef[3], Sohail Safdar[3], Zeyad T. Sharef[2], Fahad Layth Malallah[4]

[1]MODERN COLLEGE OF BUSINESS AND SCIENCE, MUSCAT, OMAN [2]FACULTY OF ENGINEERING, UNIVERSITY OF AUCKLAND, AUCKLAND, NEW ZEALAND [3]COLLEGE OF INFORMATION TECHNOLOGY, AHLIA UNIVERSITY, MANAMA, BAHRAIN [4]COMPUTER AND INFORMATION TECHNOLOGY, COLLAGE OF ELECTRONIC AND ELECTRICAL ENGINEERING UNIVERSITY, BAGHDAD, IRAQ

6.1 Introduction

Mobile ad hoc network (MANET) is a self-organized wireless mobile nodes-creating temporary network that may change its topology efficiently and randomly [1]. It is an infrastructure-less network that has no fixed routers; all nodes are capable of movement as any node can leave and join freely [2] and can be connected dynamically [1,3]. Mobile nodes communicate directly using Wi-Fi connections and cellular or satellite transmissions. These nodes are linked by wireless local area networks [4]. Each one acts as a router and host [2,5,6] and forwards packets to each other independently without any centralized administration in the network. Fig. 6–1 shows an example of mobile nodes in MANET.

Nodes in MANET communicate directly without any access point. They act as a router to provide connectivity by forwarding data packets among intermediate node to the destination. Routing protocol is used to ensure their connectivity and subsequent communication. An example of MANET communication in shown in Fig. 6–2, illustrating that each node in MANET has established the communication, such as discovering the route and forwarding the messages, by themselves. Sender node A sends a packet using a wireless channel, for example, radio signal to node B. Node A cannot reach B directly so the request will transfer through an intermediate network. Thus the packet is transmitted via hop by hop to reach the destination with the help of neighboring nodes.

MANET has wide applications [7,8] in the domain of disaster, military zones, sensor networks, emergency areas, educational aspects (seminar, workshop, and training), vehicular

Swarm Intelligence for Resource Management in Internet of Things. DOI: https://doi.org/10.1016/B978-0-12-818287-1.00012-7

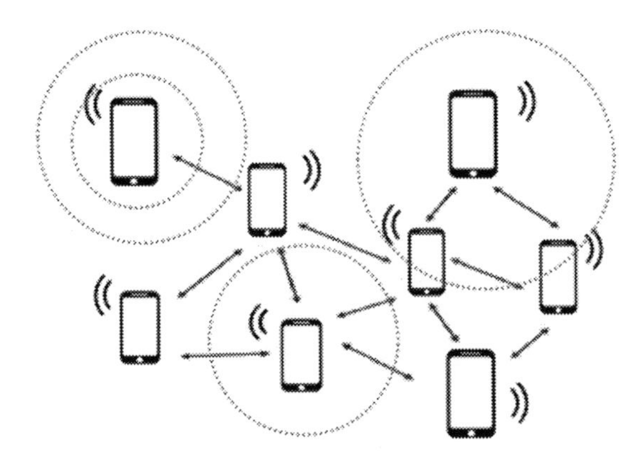

FIGURE 6–1 An example on MANET.

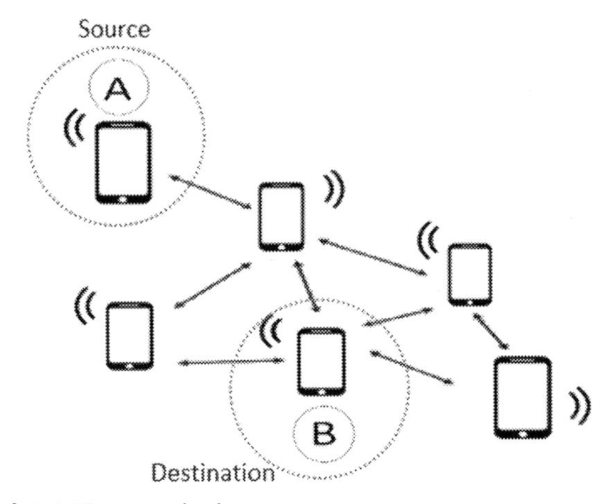

FIGURE 6–2 An example of MANET communication.

operations on the highway, entertainment, and other temporary scenarios where no physical infrastructure exists or is difficult to establish. This type of communication is more complex when compared to the infrastructure-based networks. Hence, there are numerous vulnerabilities and challenges faced by MANET due to its characteristics. The main objective of this research is to critically review and highlight the vulnerabilities within the routing protocols that lead to insecure MANET communication. The types of attacks are also reviewed and elaborated. Moreover, the research focuses on highlighting the existing secure mechanisms that are useful against routing attacks.

The rest of the paper is organized as follows: Section 6.2 discusses the advantages and drawbacks of MANET characteristics. Section 6.3 presents routing protocol classification with different examples, while MANET vulnerabilities are summarized in Section 6.4. Section 6.5 shows the current challenges in MANET environment. Classification of routing attacks in

MANET with some examples is explained in Section 6.6. Section 6.7 includes the proposed security mechanisms against routing attacks. This chapter presents brief conclusions in Section 6.8.

6.2 Pros and cons of mobile ad hoc network characteristics

MANET consists of directional (point-to-point) and multiple hop wireless communication nodes. Neighbor nodes communicate directly or reach other via a route [9,10]. MANET has its own characteristics that are summarized as follows [5,11,12]:

1. *Dynamic topology*: no infrastructure due to moving nodes.
2. *Distributed operation*: the control of the network is distributed among the nodes.
3. *Mobility*: nodes are freely to move in the network randomly.
4. *Multihop routing*: nodes act as router and forward packet to others.
5. *Unreliable links*: because the topology changes rapidly.
6. *Independent infrastructure less*: nodes are independent with no central controller.
6. *Limited resources*: battery life, storage, and bandwidth.
8. *Network scalability*: many issues to implement such a network, such as location management, routing, addressing, security, and height capacity wireless technology.

The strength of MANET characteristics is that it has the advantages of being able to be applied without requiring a base station. An ad hoc network is easy and fast to set up without the need to pull the cables through walls or roofs, and it can easily be extended to places. In addition, it is more flexible and adapts easily to changes in its configuration. While it is multipath, it leads to an increase in reliability. MANET fits the use of 4G features and services, with any time anywhere communication from mobile receivers. Conversely, its ad hoc characteristics also possess some problems; a major disadvantage of MANET is the dynamic topology, which affects the routing table to change the topology and routing algorithm [13]. Furthermore, the routing table update leads to increase in the routing overhead of the dynamic location change of nodes. Moreover, the freely join and leave nodes in MANET lead to continuous change in the transmission range interference which in turns leads to unstable connectivity.

6.3 Routing protocol in mobile ad hoc network

Routing protocol is a set of governing rules to find the optimal routes between mobile nodes in MANET [14]. In MANET, each mobile node acts as both a host and router. The main aim of these routing protocols is to find the shortest path. MANET needs a specific routing protocol to handle its functionality due to its characteristics, which is dynamic topology and links break with nodes mobility.

The routing protocol specifies the technique of how the routing table is formed to maintain information about its linking node, new node, and neighbors for sending a message

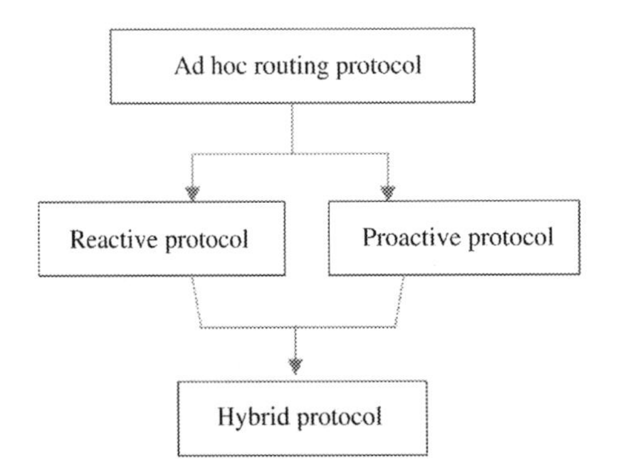

FIGURE 6–3 MANET routing protocols classification.

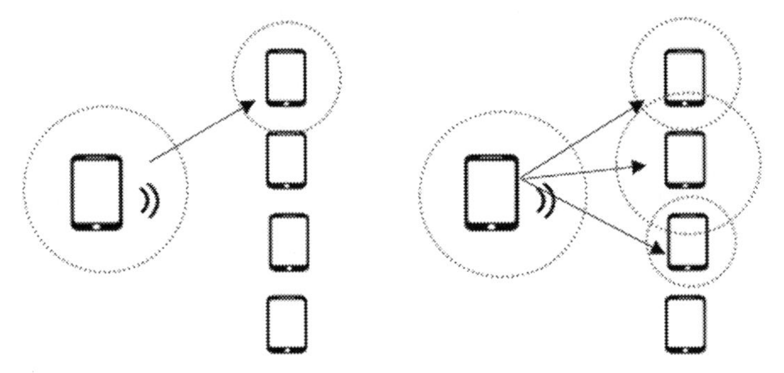

FIGURE 6–4 Casting protocols.

from a sender to a destination [7]. There are three types of routing protocols, namely, proactive, reactive, and hybrid, as shown in Fig. 6–3.

Under this classification of MANET, routing protocols can be unicast and multicast. A unicast routing protocol is a protocol that is based on one-to-one communication between one source node and one destination node. This is the most common operation in a network, for example, ad hoc on demand distance vector (AODV). While a multicast routing protocol is a protocol that is based on one to many communications between one source node and many destination node. This protocol constructs a desirable routing tree or mesh [7,15], for example, ad hoc on demand multicast distance vector. Fig. 6–4 shows the differences between unicast and multicast.

The MANET protocol stack is mostly similar to the TCP/IP model [15]. Fig. 6–5 illustrates the similarity and distinction between them. Note that in the MANET protocol stack, the ad hoc routing protocol is in the network layer.

6.3.1 Table-driven proactive routing protocols

This baseline algorithm works on a distance-vector routing and link-state routing by using a routing table to maintain and update each node frequently [16]. The routing protocol creates

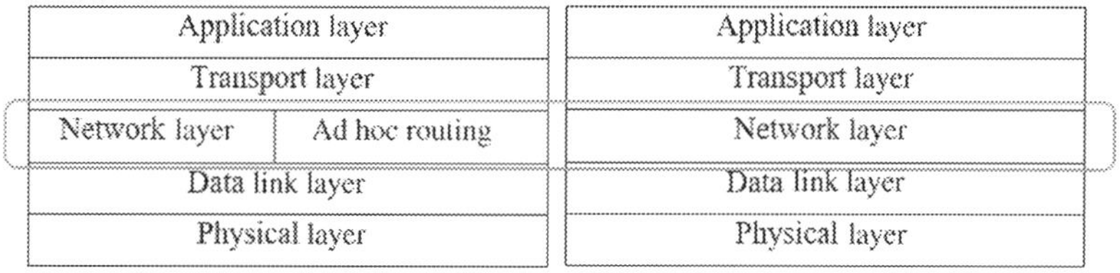

Application layer		Application layer
Transport layer		Transport layer
Network layer	Ad hoc routing	Network layer
Data link layer		Data link layer
Physical layer		Physical layer

FIGURE 6–5 MANET versus TCP/IP protocol stack.

Table 6–1 Summary of MANET routing protocols approach and features.

Routing protocol	Approach	Feature	Advantages	Disadvantages
Proactive	Use distance vector routing and link-state	• Flat/Hierarchical organization • Route always available • Suitable for LAN	• Increase route determination • Less delay • Low latency	• Overhead high • Storage requirement increase • High traffic
Reactive	Use on-demand routing protocol	• Flat organization • Route available when need • Low scalabilities	• Overhead low • Lower traffic	• Not optimal bandwidth utilization • High delay • High latency
Hybrid	Use both	• Hierarchical organization • Use both to set up route • Suitable for large network	• Overhead medium • Increase rout discovery • Lower latency in intrazone • Lowest traffic	• High latency in interzone

a routing table that frequently maintains the updated state of the network topology and has the routing information available before it is needed and for that, they are called the table-driven protocols [10]. Each node updates the routing tables and exchanges frequently the topological information with the neighboring nodes in order to maintain a consistent network. The main advantages of this protocol is that the path already exists, so the connection time is very fast and consequently, the routing latency is reduced. There is a drawback that it is very expensive for power-constrained environments as routing information must be flooded in the network frequently and has a low reaction to topology changes. Examples of this routing protocol are destination sequenced distance vector (DSDV) and optimal link state routing (OLSR) [15,16].

6.3.2 On-demand reactive routing protocols

This algorithm does not provide the routing table updates frequently. It only floods the network with route request (RREQ) packets when needed [17]. The strength of this reactive protocol is that it reduces the traffic when needed for routing and route discovery eliminates the

Table 6–2 Summary of routing attacks and countermeasures.

Types of attacks	Countermeasure
Flooding aims to consume network resources and interrupt routing setup via using HELLO massage	• When a node exceeds the predefined fixed threshold (RREQ rate) then the ID of nodes records it in blacklist • Similar to first one but it uses a statistical analysis of RREQs
Blackhole aims to cause all neighbor nodes to route packets towards it by using fake RRPE	• Uses route confirmation request (CREQ) and route confirmation reply (CREP) to avoid the Blackhole attack • Source node has to wait until receiving of the RREP packet from more than two nodes to judge the rout is safe • Considers a node is malicious if destination sequence number is arbitrarily high
Link spoofing aims to gain access to the network through false ID to misguide other nodes to create routes towards itself	• Location information deduction by using cryptography with a GPS and a time • Adding two-hop information to a HELLO message
Replay aims to inject and disturb network performance via record other node control message as it gain access to data	• Uses a time stamp with the use of an asymmetric key
Wormhole aims to drops data traffic and prevent discovery route by fake RREP using tunneling	• Multipath routing (SAM) to deduct this attack • Uses temporal leashes and geographical leashes • Similar to the geographic leashes technique as well as based on location and time-stamp synchronization
Colluding misrelay aims to alter data packets through disruption of routing operation	• Uses a conventional acknowledgment-based approach
Rushing aims to prevent discovery route via use of the tunnel procedure to form a wormhole	• Uses set of generic mechanisms together, secure neighbor detection, secure route delegation, and randomized RREQ
Sinkhole aims to drop all the data packets through incorrect route information	• Uses secure neighbor detection approach to verify that the other is within a given maximum transmission range
Sybil aims to distributed network performance and prevent fair resources as it acts as several different identities/ nodes rather than one	• Uses distributed technique to detect concurrent use of all the identities

need to keep a list of all links. Unlike proactive protocol, reactive protocol increases latency because of the additional step of route discovery and introduces a delay when a route is not found. Examples of this routing protocol are AODV [18,19] and dynamic source routing (DSR) protocol [19,20].

6.3.3 Hybrid routing protocols

It is a combination of proactive and reactive routing protocols. These are designed to increase scalability and performance via allowing nodes to form a backbone in order to

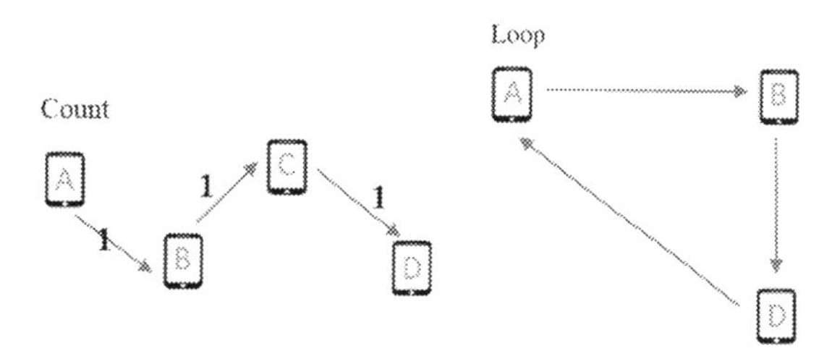

FIGURE 6–6 Diagram illustrating loop problem and count to infinite problem.

reduce the route discovery overheads [21]. At the beginning, it behaves like proactive routing protocol in which, initial nodes have a routing table. Then whenever nodes find that they do not have routes to target, they start route discovery and behave like reactive routing protocols [22]. The main advantages are the reduction of overheads when compared to pure proactive protocols and fast routing discovery by reducing delays related to reactive protocols. An example of this routing protocol is zone routing protocol (ZRP) [23,24] (Table 6–1).

The following sections show some examples of routing protocols in the three classifications.

6.3.3.1 Destination sequenced distance vector protocol

In this protocol, each node contains a routing table, which maintains destination, number of hop, and sequence number. The update of routing table occurs instantly once any alteration occurs in the network [23]. Advertisement of topology changes can be done in both ways by broadcasting and multicasting to existing neighbors. The advantages of DSDV are fresh routes are specified by sequence numbers. In addition, it maintains a single best routes to a destination, by selecting a distance vector routing in the shortest path routing algorithm. It also guarantees loop free routes and reduces the count to infinity problem. Fig. 6–6 represents the interpretation of these important concepts. The drawbacks include high routing overhead and low throughput as it frequently updates the routing table and ultimately consumes resources (e.g., bandwidth). Furthermore, it does not support multipath routing.

6.3.3.2 Optimal link state routing

This protocol is based on multiple point relay (MPR) technique in the network. When the source node sends the data to a destination node, data is rotated to multiple points with a HELLO message [15,16]. Fig. 6–7 shows an example of MPR where node A is having two MPR Q and N. Nodes exchange the routing table to know their two-hop neighbor list and choose the multipoint relay. Q and N forward the packet to A. The main advantage of OLSR is that it is minimizing the size of the control packet and produces high throughput while minimizing packet drop. However, the weaknesses are bandwidth wastage, lack of 2-hop neighbor knowledge, consumption of resources due to regular updates and high control overhead.

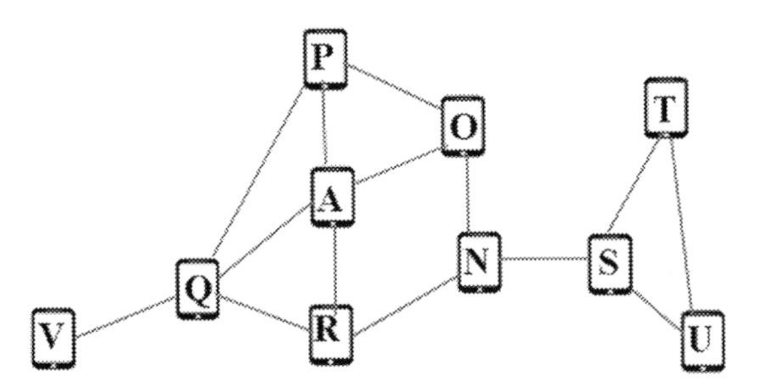

FIGURE 6–7 OLSR example.

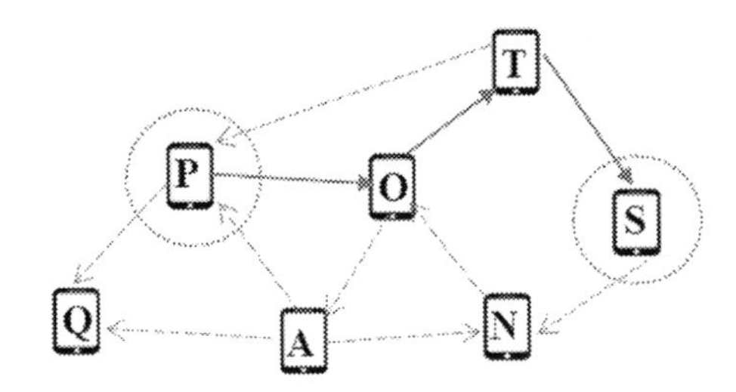

FIGURE 6–8 AODV example of RREQ.

6.3.3.3 Ad hoc on-demand distance vector protocol

AODV is an upgraded form of DSDV [24]. AODV protocol uses message mechanisms such as RREQ, route reply (RREP), and route error [20].

When a node wants to create a route, it uses a route discovery process that can be done through flooding RREQ, as shown in Fig. 6–8. All neighbors are included in this process as they receive RREQ messages. After reaching the destination RREP, as shown in Fig. 6–9, is sent back through the same path as used by the RREQ packet. During this communication, each node maintains a sequence number and routing table. The loop-free operation helps in repairing the broken links [25]. AODV is mostly capable of unicast, broadcast, and multicast routing [26]. The advantages of AODV are the use of sequence number to find fresh route, route established on request, reduction of overhead as control, and operation performed is loop free. However, the drawbacks are route setup latency when a unique route is requested, effect on loss of throughput in high mobility scenario, and packet loss due to unstable route selection.

6.3.3.4 Dynamic source routing protocol

This protocol is designed for the dynamic multihop networks based on demand routing protocol. It works in two steps, that is, route discovery and route maintenance [20,25]. When a source node wants to send a packet, it first checks the route of the node in a cache, if not found it uses route discovery. Once the packet reaches the destination, the destination node stores the information contained by RREQ and sends a reply back to the sender, as shown in

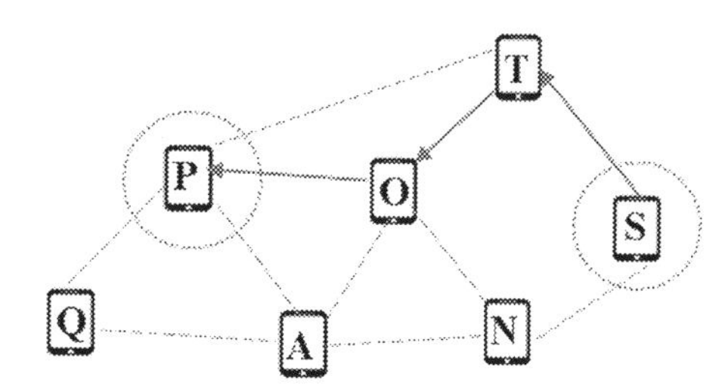

FIGURE 6–9 AODV example of RREP.

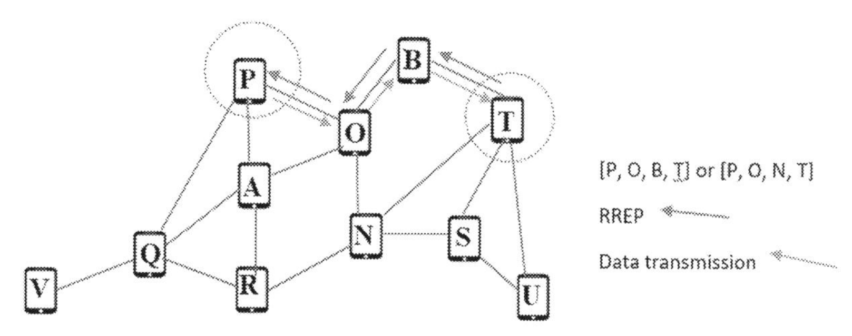

FIGURE 6–10 DSR example.

Fig. 6–10. It illustrates the steps for DSR that start with route discovery when the path is not found. Once the path is set up, it sends RREP from destination T to source node P. After P receives the RREP including route destination, data is transmitted using intermediate nodes. An advantage of DSR protocol is that it does not cause flooding to the network by updating the routing table. Furthermore, it avoids routing loops when no traffic occurs and there is no bandwidth wastage, whereas the drawback is the high delay in connection setup. Also the routing overhead increases and it is not effective in traffic-intensive and large networks.

6.3.3.5 Zone routing protocol

This protocol uses the strength of both reactive and proactive protocols to obtain better results [27]. It exhibits the overhead reduction that exists in the proactive approach and reduces the latency for discovering a new route as experienced in the reactive approach. ZRP maintains routers to all nodes within the source node's zone.

ZRP routing protocols consist of different modules, as shown in Fig. 6–11 [22].

- *Intrazone routing protocol*: this protocol is taken from the table-driven protocols that use only the local topology range which works within the specified zone.
- *Interzone routing protocol*: this protocol is taken from the domain that is used when the route among different zones is needed for the communication between the source and destination.

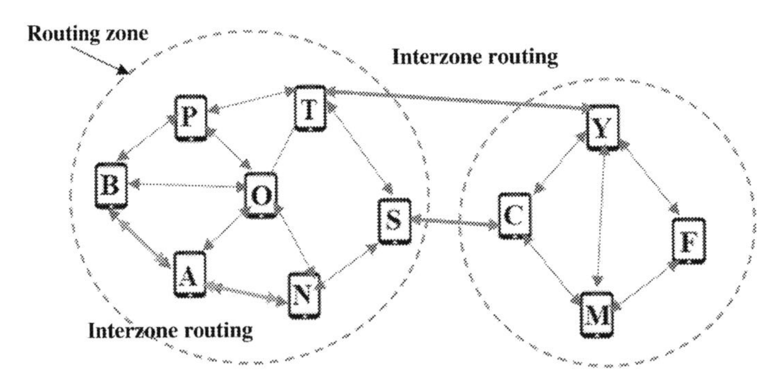

FIGURE 6–11 ZPP example.

The main advantage gained from ZRP is the lower control overhead, but it doesn't provide an optimized shortest path.

6.4 Mobile ad hoc network vulnerabilities

Vulnerability is a weakness that can be exploited by a security breacher. In MANET, there are various security-related vulnerabilities. Since any nodes can join MANET without verifying their identity, this results in potential unauthorized access. This concern may be considered as a vulnerability to unofficial data manipulation. Comparing to wired networks, MANET is considered much more vulnerable [28]. Some of the vulnerabilities are listed as follows:

- *Lack of centralized management*: in a mobile ad hoc network, there is no central monitoring unit, hence it causes difficulty in detecting attacks and monitoring the traffic in a large-scale dynamic network [29]. This leads to a lack of the secure trust management amongst nodes while they communicate.
- *Resource availability*: this is considered as major issue in MANET since it requires various resources [5,30]. Providing secure communication and protection against threats in a dynamic environment leads to the development of various security schemes. There are some constraints which make nodes vulnerable and alluring for attackers, which are limited bandwidth, high cost, slow links, and power constraints [31].
- *Dynamic topology*: MANET refers to the concept of changeable/mobile nodes. This results in disturbing the trust relationship among nodes [32]. In addition, if malicious nodes are detected on the network, this result in disruption of services to other nodes since the trust is violated. These nodes mobility could be better protected with a distributed and adaptive security mechanism [29].
- *Scalability*: MANET is constantly scalable due to nodes' mobility which impacts the security due to its expansion [28,29]. Security mechanisms have to be applied to handle different types of network scales.

- *Cooperativeness*: MANET nodes are usually cooperative as nonmalicious according to the routing algorithm. Hence, a malicious node can easily act as a routing agent and interrupt the network performance as it changes the protocol specification [13,31].
- *Limited power supply*: power consumption is related to the operational lifetime of a mobile ad hoc network. Because of that, reducing the power consumption is considered a major challenge as the battery in MANET represents independent nodes [28]. Moreover, a node in MANET may behave in a selfish manner when a power supply is limited. End of power battery can result to increase the probability of network dysconnectivity over MANET and this effects the node communication [31].
- *Bandwidth constrain*: MANET, as a wireless network, has limited links capacity compared to a wired network [13]. Mobile nodes are more vulnerable to environmental disturbances, external noise, interference, and signal attenuation effects that can degrade the quality of the network services.
- *Adversary inside the network*: mobiles nodes are free to leave and join the network. Therefore they may behave maliciously if infected or malfunctioned, which is considered more dangerous than the deliberate external attack on a specific node [29].

6.5 Mobile ad hoc network challenges

The list below shows the current challenges in MANET environments that must be addressed carefully before a wide commercial deployment can be made.

- *Routing overhead*: one of the challenging tasks in MANET is the constant change in the network topology where nodes move fast and subjectively. This causes an issue in the routing packet between any pair of nodes as well as unnecessary routing overhead [5,16].
- *Quality of service (QoS)*: it is difficult to fulfill QoS in a constantly changing environment due to the inherent stochastic feature of communication quality in MANET. QoS must be implemented in order to maintain the best effort of service for the multimedia communication services and voices [16].
- *Security and reliability*: mobile ad hoc networks are naturally exposed to numerous security attacks. Security is one of the challenges that should be considered to ensure that the data is transferred safely and completely. The MANET system security can be enhanced through the least privilege principle and hybrid models [33]. In addition, the distributed operation feature requires different schemes of authentication and key management. Furthermore, limited transmission range leads to reliability issues, hidden terminal problems, mobility-induced packet losses and data transmission errors [34].
- *Interworking*: communication between ad hoc and infrastructure networks is often expected in different cases. The coexistence of mobile routing protocols is a challenge for harmonious mobility management [33].
- *Multicast*: multicast supports multiple members in wireless communication where nodes need to form collaborative working groups using the broadcast performances of wireless communication and the limited wireless channel resources effectively. The multicast

directing convention must be equipped with a multicast enrollment element (leave and join) which does not remain static for a long time [27].

- *Power consumption*: power preservation and power-aware routing must be taken into account. In lightweight mobile terminals, the communication functions should be enhanced for less power utilization.
- *Location aided routing*: the use of position information helps in increasing the routing protocol performance and finding new routes, which will subsequently reduce the number of routing messages significantly [27].
- *Packet losses due to transmission errors*: there are many reasons that lead to high risk of packet losses. For example, the increased collision caused by the presence of hidden terminals, common path breaks due to nodes mobility, and unidirectional links.
- *Security threats*: based on MANET characteristics and vulnerabilities there are two types of attacks, that is, passive and active, which are introduced in the following section. It leads us to conclude that security is a major challenge in this type of network.

6.6 Routing attacks in mobile ad hoc network

Classification of routing attacks in MANET depends on the malicious behavior of a node on network. The malicious node is used to either read the secret information or change the information. Therefore the attacks are classified into two categories [35].

6.6.1 Passive attack

A passive attack does not change the data transmitted within the network or disrupt the operation of a routing protocol. However, it includes the unauthorized "listening" to discover the important data from routed traffic.

6.6.2 Active attack

Active attacks attempt to change or destroy the data in the network. These attacks can be internal or external, Fig. 6–12 shows the difference between them.

- *Internal attacks*: these attacks are carried out by compromised nodes that are part of the network. Malicious nodes change the routing information by advertising themselves as having the shortest path to the destination.
- *External attacks*: these attacks are carried out by nodes that do not belong to the network. These malicious nodes create additional overhead, drop data packets, cause denial of services (DoS), and advertise wrong routing information.

Attacks in different layers of MANET are dealt with in the same way as for an infrastructure network except for the network layer, where routing protocols are more vulnerable to attack because of the cooperative nature of the nodes and the lack of infrastructure for routing. Thus routing protocols need to be protected against threats and attacks. The malicious

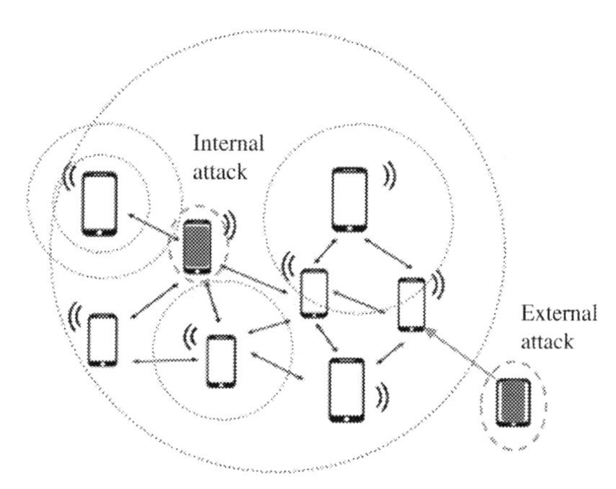

FIGURE 6–12 Difference between internal and external attack in MANET.

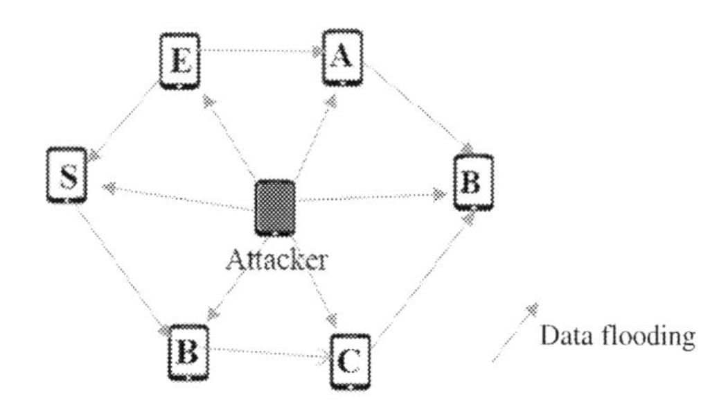

FIGURE 6–13 Example of flooding attack in MANET.

node(s) can attack MANET using different methods, such as sending fake messages, fake routing information, and fake advertising links to disrupt the routing operations. For that reason, many security techniques and mechanisms have been proposed to prevent malicious behavior, but none of those preventive approaches is capable of defending against all attacks [36].

The following parts of this paper explain the current routing attacks in MANET and the solutions that are proposed to counter those routing attacks.

- *Flooding attack*: these attacks work either by using RREQs or data flooding. The aim of it is to consume network resources and interrupt the routing setup to reduce network performance [37]. For example, in AODV protocol, a malicious node sends a huge number of RREQs for nonexisting destinations in a short period. Because there are no replies, the requests will flood the whole network, as shown in Fig. 6–13. As a result, power and bandwidth of all the nodes will be consumed, which could lead to service denial.
- *Black hole attack*: a malicious node aims to halt other nodes' routing data packets through fake route information as if it is an ideal route. Then it drops all packets instead

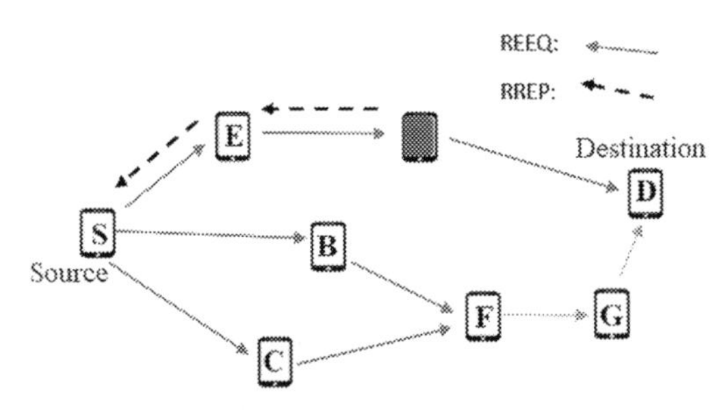

FIGURE 6–14 Example of black hole attack in MANET.

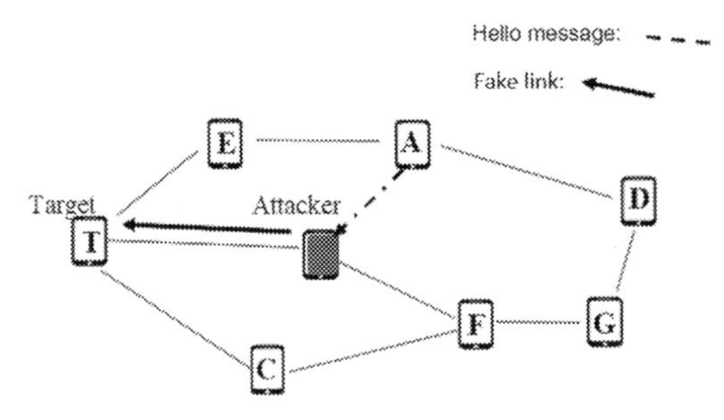

FIGURE 6–15 Example of link spoofing attack in MANET.

of forwarding them [38]. For example, in AODV protocol, the attacker sends fake RREP to the source showing a sufficient fresh route. This causes the source node to select this route; thus all traffic will be routed to the attacker. Fig. 6–14 shows the flow of a black hole attack in MANET.

- *Link withholding and spoofing attack*: this attack aims to disconnect links among nodes. This is done via a malicious node that blocks link broadcasts of a specific node or a group of nodes. However, in spoofing attack a malicious node broadcasts fake route information to disrupt the routing operation [39]. As a consequence, the malicious node manipulates the data or routes traffic. For example, in the OLSR protocol, as shown in Fig. 6–15, an attacker broadcasts a fake link targeting two-hop neighbors. The target nodes select the malicious node to be its MPR. The malicious node thereby can modify or drop the routing traffic and this may lead to DoS.
- *Replay attack*: these attacks occur when a node recodes another node's valid message control [40]. A malicious node recodes control messages of other nodes and resends them later. This forces the nodes to record their routing table with stale routes. The attacker repeatedly retransmits the valid data into the network and disturbs the routing operation.
- *Wormhole attack*: it is one of the most aggressive attacks in MANET. In this attack, wormhole nodes create a fake route as if it has the shortest path to destination. It uses tunneling between two or more malicious nodes that are contributing in this attack. The

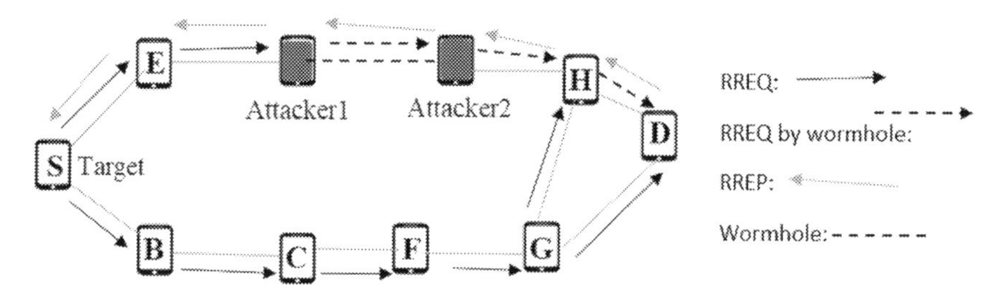

FIGURE 6–16 Example of wormhole attack in MANET.

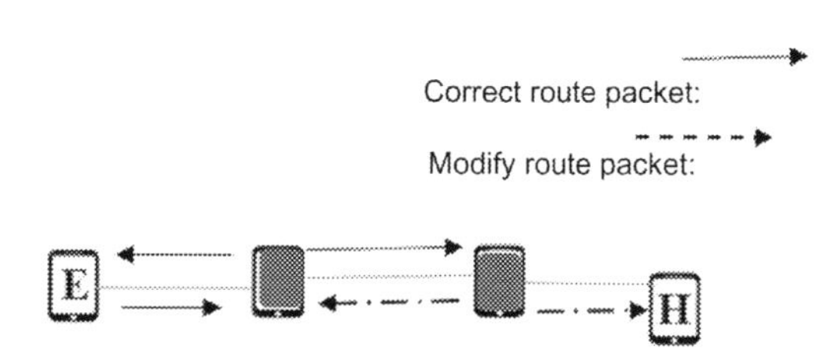

FIGURE 6–17 Example of colluding misrelay attack in MANET.

tunnel here is known as a wormhole, as shown in Fig. 6–16. For example, in DSR and AODV routing protocols, the attack could prevent the discovery of any routes through the wormhole if there is no defense mechanism in these routing protocols. As a result, they will not be able to discover any valid routes [41].

- *Colluding misrelay attack*: it occurs when many malicious nodes work in collusion to alter or drop data packets to interrupt routing operation in MANET. This attack is difficult to detect through the conventional methods, such as watchdog and pathrater [39]. For example, an attack occurs in OLSR protocol where two malicious nodes are present, as shown in Fig. 6–17. One attacker forwards routing packets to avoid reaching their destination and the second attacker drops or modifies the routing packets.
- *Rushing attack*: this type of attack mainly strikes the on-demand routing protocols and sends a route discovery packet to the target nodes [38]. They use the tunnel procedure to form a wormhole between two of them. For example, a malicious node sends RREQs for the discovery and forwards them to the neighbor, destination, or target node. Then, the attacker includes hops that discover the route by discovery route. As a result, the target receives false requests from the malicious node, then it forwards the request to other nodes. The request will no longer be forwarded because of the node capability and the RREQs from the original node will be ignored. As a result, the initiator will not be able to discover any usable routes.
- *Sinkhole attack*: these are categorized as route interference attacks [38]. In this attack, malicious nodes send incorrect route information to the neighboring nodes suggesting that there is a favorable low-cost route available close to the destination. As a result, the neighboring nodes send all the packets to the malicious nodes that ultimately draw all the traffic, or part of it, to themselves.

- *Sybil attack*: a malicious node produces itself as a large number of nodes instead of a single node. Sybil attacker creates incorrect identities of a number of additional nodes which are called Sybil nodes [40]. They fabricate a new identity for themselves or they steal an identity of the legitimate nodes. Sybil nodes may affect routing protocol operation by placing themselves at different locations. This makes it difficult to identify a maliciously behaving node and prevents fair resource allocation among the nodes.

6.7 Proposed security mechanisms applied against routing attack

The following are the existing security solutions to overcome the abovementioned routing attacks.

- *Security mechanisms against flooding attack*: Yi Ping et al. have proposed a mechanism to prevent a flooding attack in the AODV protocol. The mechanism works such that, when a node exceeds the predefined threshold (RREQ rate) then the ID of nodes is recorded in blacklist. Afterwards, all RREQs from the blacklisted nodes will be dropped. The drawback is that a flooding threshold has to be set, and it may lead to false alarms. Furthermore, a malicious node impersonates the ID of a genuine node and broadcasts a large number of RREQs. As a result, it might put the ID of this genuine node on the blacklist by mistake [42,43]. Wu et al. have proposed a technique which is similar to the first one but instead of using a fixed threshold, it uses a statistical analysis of RREQs in order to detect malicious RREQ floods. The main advantage of this approach is that it can decrease the influence of the attack for varying flooding rates [40,44].
- *Security mechanisms against blackhole attack*: Lee et al. have introduced route confirmation request (CREQ) and route confirmation reply (CREP) to avoid the blackhole attack. In this approach, the intermediate node sends RREPs to the source node as well as sending CREQs to its next-hop node to reach the destination. The next-hop node receives CREQ then it will screen its cache for a route to the destination. If it has the route in cache, it sends the CREP to the source. After that, source node compares CREP and RREP. If both match, the source node assumes that the path is optimal. One disadvantage of this technique is that it cannot avoid an attack in which two sequential nodes work in collusion, because the next-hop node is a colluding attacker that is sending (CREP) to support the incorrect route [39,45]. Shurman et al. have proposed that the source node has to wait until receiving the RREP packet from more than two nodes. Then, it checks a shared hop from receiving RREPs and judges that the route is safe if at least one hop is shared. The drawback of this technique is the time delay, because the source node must wait until multiple RREPs arrive [42,46]. Chavda suggested countermeasure for detection of malicious nodes if the destination sequence number is arbitrarily high based on differences between the destination sequence numbers of the received RREPs. The feature of this countermeasure is that it can detect the attack at a low cost without modification of the existing protocol or introducing extra routing [43,47].

- *Security mechanisms against spoofing attacks*: Raffo et al. have proposed a technique to prevent this attack based on location information deduction by using cryptography with GPS and time. Each node needs to advertise the location of itself by the GPS and the time stamp in order for nodes to get location information of each other. By calculating the distance between two nodes based on a maximum transmission range link spoofing can be deducted. The weakness of this mechanism is that it might not work if all nodes are not associated with a GPS, here a malicious node can still advertise fake information and make it hard to be deducted [41,48]. Kannhavong et al. have proposed a solution to detect the link spoofing attack by adding two-hop information to a HELLO message. Each node advertise its two-hop to neighbors to enable each node to learn the complete topology up to three hops in order to detect the inconsistency when the link spoofing attack is launched. This method can detect a link spoofing attack without using special hardware such as a GPS or time stamp synchronization, which is considered as main advantage. On other hand, the limitation is that this technique might not detect link spoofing with nodes further away than three hops [41,49].
- *Security mechanisms against replay attack*: a solution is proposed by using a time stamp with the use of an asymmetric key. It prevents this attack by comparing the current time and time stamp contained in the received message. If time stamp is too far from the current time, the message will consider as malicious and it's refused [40,46].
- *Security mechanisms against wormhole attack*: Qian et al. have proposed multipath routing (SAM) to deduct this attack. It calculates the relative frequency of each link appearing in obtained routes from one route discovery, the highest one is identified as a worm-hole attack. The advantage of the SAM technique is that it introduces limited processing overhead. On the other hand, it might not work in a nonmultipath routing protocol, such as a pure AODV protocol [42,50]. Another approach proposed that packet leashes detect and defend against the wormhole attack. There are two types of leashes: temporal leashes and geographical leashes. For the temporal leash approach, a packet is controlled from traveling further than a specific distance as each node computes the current time of the packet and packets include the expiration time. The receiver of the packet checks if the packet expires or not by comparing its current time and the expiration time. The main drawback is that it requires all nodes to have tightly synchronized clocks. For the geographical leash, each node is required to know its own position and have loosely synchronized clocks. A sender includes its current position and the sending time of a packet. A receiver computes distance between itself and the sender of the packet to check neighbor relations. The advantage of geographic leashes over temporal leashes is that the time synchronization does not need to be very tight [40,49]. Another work proposed a method against wormhole attack in the OLSR protocol is similar to the geographic leashes technique. This technique is based on location and time-stamp synchronization between all nodes. A sender of a HELLO message includes the current time plus adds 0.3 random seconds of delay along with its position. When a neighbor receives a HELLO message, it calculates the difference between arrival times of HELLO messages and the included one in the hello message timing profile and uses it for

the detection of attacker nodes. Unaffected nodes should have frequently larger intervals between the data packets [51,52].

- *Security mechanisms against colluding misrelay attack*: this type of attack might be detected via conventional acknowledgment-based approach. This approach requires all nodes to return an acknowledgment, that is, ACK, but this leads to a large overhead on the network, which is considered to be inefficient. Another proposed method to detect an attack is that when multiple malicious nodes attempt to drop packets, they are required to increase their transmission power twice, as a result the attacker can be detected. However, this approach might not detect the attack in which more than two colluding attackers work in collusion [40].
- *Security mechanisms against rushing attack*: Hu et al. have proposed a technique to defend against the rushing attack. This approach consists of a set of generic mechanisms together: secure neighbor detection, secure route delegation, and randomized (RREQ) forwarding to each node. To verify that the other nodes are within the given maximum transmission range or not, a secure neighbor detection is used. Two nodes A and B should be in allowable transmission range, then a node A determines that node B is a neighbor, it signs a route delegation message which allows node B to forward the RREQ. Moreover, node B signs an accept delegation message if node A is in the allowable transmission range too. This mechanism replaces the traditional duplicate suppression in on-demand route discovery with the randomized selection of RREQ [42,50].
- *Security mechanisms against sinkhole attack*: Hu proposed secure neighbor detection that allows each neighbor to verify that the other is within a given maximum transmission range, and secure route delegation and randomized RREQ forwarding replaces traditional duplicate suppression in on-demand route discovery [51,53].
- *Security mechanisms against Sybil attack*: Abbas has proposed a distribution-based technique for Sybill attack detection in case of the concurrent use of all identities by attackers. Sybil attack can be indicated by capturing the movement of all identities. If identities traveling in the same path, this considered as Sybil attack. The disadvantage that it requires directional GPS for location determination [40,54] (Table 6–2).

6.8 Conclusion

This paper presented a review of the MANET concept in ad hoc basis mode. In addition, it discussed the most significant characteristics of MANET and the relevant pros and cons. MANET due to its characteristics is more vulnerable to attacks. In addition, this paper covers MANET routing protocols, its vulnerabilities, and different types of attacks. Furthermore, it explains the proposed security mechanisms against routing protocol attacks. MANET is considers to be a low-cost communication network compared with other infrastructure networks. Nowadays, use of this network is very common as it trends towards large scale and mesh architecture. It is highly recommended to overcome the recent challenges and weaknesses to provide a trustworthy and highly secure level of communication. One of these required features is the improvement in capacity and bandwidth, which involves the demand of enhanced spatial spectral reuse and higher frequency.

References

[1] L. Raja, S. Baboo, An overview of MANET: applications, attacks and challenges, Int. J. Comput. Sci. Mob. Comput 3 (1) (2014) 408–417.

[2] N. Raj, P. Bharti, S. Thakur, Vulnerabilities, challenges and threats in securing mobile ad-hoc network, in: Proc. – 2015 5th Int. Conf. Commun. Syst. Netw. Technol. CSNT 2015, pp. 771–775, 2015.

[3] T. Kunz, B. Esfandiari, F. Ockenfeld, Efficient routing in mobile ad-hoc social networks, in: Proc. 2017 IEEE Int. Conf. Internet Things, IEEE Green Comput. Commun. IEEE Cyber, Phys. Soc. Comput. IEEE Smart Data, iThings-GreenCom-CPSCom-SmartData 2017, vol. 2018–January, pp. 216–222, 2018.

[4] B.T. Sharef, R.A. Alsaqour, M. Ismail, Vehicular communication ad hoc routing protocols: a survey, J. Netw. Comput. Appl. 40 (2014) 363–396.

[5] G.S. Dhillon, Vulnerabilities & attacks in mobile adhoc networks (MANET), Int. J. Adv. Res. Comput. Sci. 8(4), 2017, 373–375. Available online at: <www.ijarcs.info>.

[6] R. Singh, An overview of MANET: characteristics, applications attacks and security parameters as well as security mechanism, Int. Res. J. Eng. Technol. 5 (9) (2018) 1155–1159.

[7] B.U. Islam, R.F. Olanrewaju, F. Anwar, M. Yaacob, A.R. Najeeb, A survey on MANETs: architecture, evolution, applications, security issues and solutions, Indones. J. Electr. Eng. Comput. Sci. 12 (2) (2019) 832.

[8] R. Gomathijayam, V. Santhi, A. Professor, A survey on issues and challenges of MANET, Int. J. Electr. Electron. Comput. Sci. Eng. Spec. Issue -ICDSSI (2018). ISSN: 2348–2273.

[9] S. Guo, Performance evaluation of {lifetime-aware} multicast routing protocols in mobile ad hoc networks, in: Fourth Int. Conf. Mob. Comput. Ubiquitous Netw., 2008.

[10] B. Sharef, R. Alsaqour, M. Alawi, M. Abdelhaq, E. Sundararajan, Robust and trust dynamic mobile gateway selection in heterogeneous VANET-UMTS network, Vehicular Commun. 12 (2018) 75–87.

[11] A. Kaid, S. Ali, U.V. Kulkarni, Characteristics, applications and challenges in mobile ad-hoc networks (MANET): overview, Wireless Networks 3 (12), 2015, 6–12.

[12] D. Ahmed, O. Khalifa, An overview of MANETs: applications, characteristics, challenges and recent issues, IJEAT 3 (4) (2017) 128.

[13] N. Raza, M.U. Aftab, M. Qasim Akbar, O. Ashraf, M. Irfan, Mobile ad-hoc networks applications and its challenges, Commun. Netw. 8 (8) (2016) 131–136.

[14] M.S.M.C., Congestion avoidance and CONTROL mechanisms for manet using aodv protocol: a survey, Int. J. Adv. Res. Comput. Sci. 9 (2), 2018, 19–24. Available online at: <www.ijarcs.info>.

[15] V. Bhatt, S. Kumar, Study and literature or research survey of routing protocols and routing attacks in MANET with different security technique in cryptography for network security, Int. J. Futur. Revolut. Comput. Sci. Commun. Eng. 4 (4) (2018) 839–845.

[16] A. Patil and G.B.Hangargi, An application, challenges and routing protocol in Mobile Ad-Hoc Network, Int. J. Ethic. Eng. Manage. Education 2 (5), 2015, 43–48.

[17] N. Raj, P. Bharti, S. Thakur, Vulnerabilities, challenges and threats in securing mobile ad-hoc network, in: Proc. – 2015 5th Int. Conf. Commun. Syst. Netw. Technol. CSNT 2015, pp. 771–775, 2015.

[18] S. Habib, S. Saleem, K.M. Saqib, Review on MANET routing protocols and challenges, in: Proc. – 2013 IEEE Stud. Conf. Res. Dev. SCOReD 2013, no. December, pp. 529–533, 2015.

[19] N. Gupta, R. Gupta, Routing protocols in Mobile Ad-Hoc Networks: an overview, in: Int. Conf. "Emerging Trends Robot. Commun. Technol. INTERACT-2010", pp. 173–177, 2010.

[20] S. Wali, S.I. Ullah, A.W.U. Khan, A. Salam, A comprehensive study on reactive and proactive routing protocols under different performance metric, Sukkur IBA J. Emerg. Technol. 1 (2), 2018, 39–51.

[21] A. Zain, H. El-khobby, H.M. Abd Elkader, M. Abdelnaby, MANETs performance analysis with dos attack at different routing protocols, Int. J. Eng. Technol. 4 (2) (2015) 390.

[22] K. Raheja, S.K. Maakar, A survey on different hybrid routing protocols of MANET, Int. J. Comput. Sci. Inf. Technol. 3 (5) (2014) 5512–5516.

[23] T. Gui, C. Ma, F. Wang and D.E. Wilkins, 2016, Survey on swarm intelligence based routing protocols for wireless sensor networks: an extensive study. In 2016 IEEE International Conference on Industrial Technology (ICIT) (pp. 1944-1949). IEEE

[24] R.K. Singh, P. Nand, Literature review of routing attacks in MANET, in: Proc. – IEEE Int. Conf. Comput. Commun. Autom. ICCCA 2016, pp. 525–530, 2017.

[25] N. Gupta, R. Gupta, Routing protocols in Mobile Ad-Hoc Networks: an overview, in: Int. Conf. "Emerging Trends Robot. Commun. Technol. INTERACT-2010, pp. 173–177, 2010.

[26] S. Habib, S. Saleem, K.M. Saqib, Review on MANET routing protocols and challenges, in: Proc. – 2013 IEEE Stud. Conf. Res. Dev. SCOReD 2013, no. December, pp. 529–533, 2015.

[27] A.S. Bundela, G. Sharma, P. Panse, S. Solanki, A secure routing in ad-hoc network, in: 2016 Symp. Colossal Data Anal. Networking, CDAN 2016, pp. 1–5, 2016.

[28] L. Raghavendar Raju, C.R.K. Reddy, A survey on routing protocols and QoS in Mobile Ad Hoc Networks (MANETs), Indian J. Sci. Technol. 10 (9) (2017) 1–8.

[29] P. Goyal, V. Parmar, R. Rishi, MANET: vulnerabilities, challenges, attacks, application, IJCEM Int. J. Comput. Eng. Manag. 11 (2011) 2230–7893.

[30] V.M. Agrawal, H. Chauhan, An overview of security issues in mobile Ad hoc network, Int. J. Comput. Eng. Sci. 1 (1) (2017) 9.

[31] R. Dey, H.N. Saha, Different routing threats and its mitigations schemes for Mobile Ad-Hoc Networks (MANETs) – a review, IPASJ Int. J. Electron. Commun. (IIJEC) 4 (3), 2016, 27–38.

[32] A. Vij and V. Sharma, 2016, April. Security issues in mobile adhoc network: a survey paper. In 2016 International Conference on Computing, Communication and Automation (ICCCA) (pp. 561–566). IEEE.

[33] S.K.S. Kaur, An overview of mobile ad hoc network: application, challenges and comparison of routing protocols, IOSR J. Comput. Eng. 11 (5) (2013) 7–11.

[34] A. Gupta, P. Verma, R.S. Sambyal, An overview of MANET: features, challenges and applications, Int. J. Sci. Res. Comput. Sci. Eng. Inf. Technol. 1 (4) (2018) 2456–3307.

[35] S. Kaushal, R. Aggarwal, A study of different types of attacks in MANET and performance analysis of AODV protocol against wormhole attack, Int. J. Adv. Res. Comput. Eng. Technol. 4 (2) (2015) 301–305.

[36] D. Khan, M. Jamil, Study of detecting and overcoming black hole attacks in MANET: a review, in: 2017 Int. Symp. Wirel. Syst. Networks, ISWSN 2017, 2018-January, pp. 1–4, 2018.

[37] K. Gupta, P.K. Mittal, An overview of security in MANET, Int. J. Adv. Res. Comput. Sci. Softw. Eng. 7 (6) (2017) 151–156.

[38] N. Khanna and M. Sachdeva, A comprehensive taxonomy of schemes to detect and mitigate blackhole attack and its variants in MANETs, Comput. Sci. Revi. 32, 2019, 24–44.

[39] M. Ngadi, R. Khokhar, S. Mandala, A review current routing attacks in mobile ad-hoc networks, Int. J. Comput. Sci. Sec. 2 (2008) 18–29.

[40] P.N. Reddy, C. Vishnuvardhan, V. Ramesh, Routing attacks in mobile ad hoc networks, IJCSMC 2 (5) (2013) 360–367.

[41] M. Sadeghi, S. Yahya, Analysis of wormhole attack on MANETs using different MANET routing protocols, in: ICUFN 2012 – 4th Int. Conf. Ubiquitous Futur. Networks, Final. Progr., pp. 301–305, 2012.

[42] N. Hussain, A. Singh and P.K. Shukla, In depth analysis of attacks & countermeasures in vehicular ad hoc network, Int. J.Software Eng. Its Appl. 10 (12), 329–368.

[43] Y. Ping, A new routing attack in mobile ad hoc networks, Int. J. Inf. Technol. 11.2 (2005) 83–94.

[44] B. Wu, et al., A survey of attacks and countermeasures in mobile ad hoc networks, Wireless/Mobile Network Security, 17, Springer, 2006.

[45] S. Lee, B. Han, M. Shin, Robust routing in wireless ad hoc networks, in: 2002 Int'l. Conf. Parallel Processing Wksps., Vancouver, Canada, August 18–21, 2002.

[46] S. Agrawal, S. Jain, S. Sharma, A survey of routing attacks and security measures in mobile ad-hoc networks, J. Comput. 3 (2011) 41–48.

[47] K. Chavda et al., Removal of black hole attack in AODV routing protocol of MANET, in: 4th ICCCNT, Tiruchengode, July 2013.

[48] D. Raffo et al., Securing OLSR using node locations, in: Proc. 2005 Euro. Wireless, Nicosia, Cyprus, April 10–13, 2005.

[49] B. Kannhavong et al., A collusion attack against OLSR-based mobile ad-hoc networks, in: IEEE Globecom' 6.

[50] Y.-C. Hu, A. Perrig, D.B. Johnson, Ariadne: a secure on-demand routing protocol for ad-hoc networks, in: Proc. MobiCom'2, Atlanta, GA, September 23–28, 2002.

[51] T. Seminar, Analysis of security attacks regarding MANET routing protocols, vol. 5, no. December, pp. 182–189, 2016.

[52] S. Desilva, R.V. Boppana, Mitigating malicious control packet floods in ad-hoc networks, in: Proc. IEEE Wireless Commun. and Networking Conf., New Orleans, LA, 2005, pp. 2112–2117, Vol. 4.

[53] Y. Hu et al., Rushing attacks and defense in wireless ad-hoc network routing protocols, in: Proceedings of the ACM Workshop on Wireless Security, SanDiego, California, pp. 30–40, September 2003.

[54] S. Abbas, et al., Lightweight sybil attack detection in MANET, IEEE Syst. J. 7 (2) (2013).

7

Swarm intelligence for intelligent transport systems: opportunities and challenges

Elezabeth Mathew

BUID, DUBAI, UNITED ARAB EMIRATES

7.1 Introduction

Many years ago, 3C (computing, communication, and control) technology was effectively used to various physical systems like defense, energy, critical infrastructure, health care, manufacturing, and transportation, and vividly improved their controllability, adaptability, autonomy, efficiency, functionality, reliability, safety, and usability [1]. These physical systems are now controlled by Internet of Things (IoT), social network, cloud computing, big data, and intelligent systems due to technical and social development. A cyberphysical system (CPS) integrates the physical, cyber, and human factors in a single framework. Cyber systems are being gradually implanted in all types of physical systems and are making the systems more intelligent, energy-efficient, and comfortable, for example, intelligent transportation systems (ITS), factories, and cities [1]. The CPSs are suitably ever-present in daily life and their complexity is growing endlessly. A CPS by itself cannot approximate the influence of humans, organization, and societies which are unpredictable, dissimilar, and complex. This implies that CPS may become useless in some situations where human interference is necessary. Fortunately, the social behavior can be attained, evaluated, and exploited to control and manage the performance of complex systems based on the worldwide Internet and big data technologies [1]. Recently, the cyberphysical social system (CPSS) has been proposed by integrating social components into CPS. This chapter presents compelling arguments in favor of new research directions in the area of ITS that is based on a CPS perspective [2].

7.2 Intelligent transport system

The territory of transport which includes aircrafting in their size, shape, destinations, and especially how these mobility vehicles deal with their customers [2]. The entire business models, processes, and procedures that use conveyance as a means to distribute goods and

Swarm Intelligence for Resource Management in Internet of Things. DOI: https://doi.org/10.1016/B978-0-12-818287-1.00013-9

people have also been changed drastically with the emergence of artificial intelligence (AI). AI integrated into software in these transport modules enhances a faster, better, and cheaper delivery, while at the same time cutting back emissions and other waste from the perception of sustainability [2].

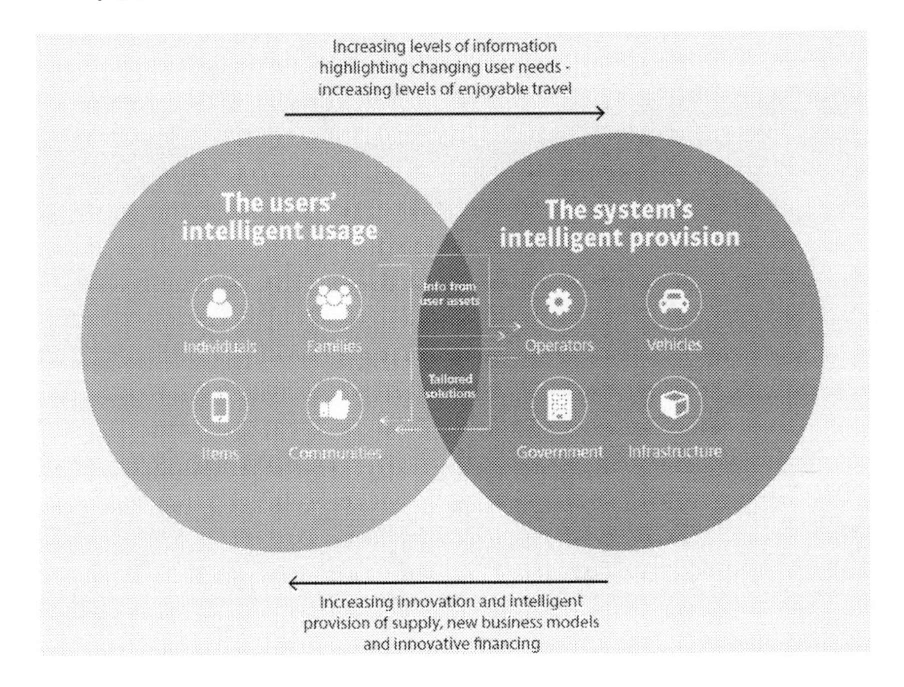

An illustration from *The journeys of the future* published by Atkins magazine is given above (2017).

7.3 Is intelligent transport system a system of systems?

A software-intensive system is, by definition, any system where software influences to a large extent the design, construction, deployment, and evolution of the system as a whole [2]. The system of systems engineering (SoSE) is an evolving interdisciplinary approach that lays an emphasis on ultralarge-scale interoperation of many independent self-contained constituent systems in order to satisfy a global need [3].

Table 7–1 explains each property of SoS and equates it with ITS. Hence, as in the concept map of SoS below, ITS depicts all the characteristics of SoS, so definitely intelligent transport system is coming under the umbrella of the acknowledged SoS [3].

The ITS has operational independence, managerial independence, is evolutionary in nature, has got emergent behavior and it is distributed geographically. Moreover, ITS belongs to an acknowledged category of SoS. Various categories of SoS are given in Table 7–2. The concept map of SoS is given below for clarification. As depicted, the concept

Table 7–1 Properties of SoS.

	Property	Proof
1	Operational independence	Each constituent system in ITS like rail, road, air, etc. can operate independently and is capable of achieving its own goals in the absence of the other systems
2	Managerial independence	The constituent systems of ITS can managed independently and can be added or removed from the SoS
3	Evolutionary in nature	The SoS adapts to fulfill its, possibly evolving, objectives as its underlying technologies and needs evolve with time
4	Emergent behavior	The functionality and behavior of the SoS develops in ways not achieved by the individual systems
5	Geographical distribution	ITS has a geographical distribution that limits the interaction of the constituent systems to information exchange

Table 7–2 ITS compared with SoS characteristics.

Categories	Directed	The interoperable SoS is built to fulfill a specific purpose
	Collaborative	The constituent systems "voluntarily" collaborate in an agnostic way to achieve an agreed-upon central purpose
	Acknowledged	There are recognized objectives, a designated manager, and resources for the SoS. The constituent systems return their independent ownership, objectives, funding, development, and sustainment approaches
	Virtual	There is no centrally agreed purpose for the SoS

of SoS contains systems that are independently managed and operated. It does not have clear boundaries and exhibits complex properties. SoS also has other properties like heterogeneity, geographical distribution, emergence, and being evolutionary in nature (Table 7–3).

Hence, as displayed in the concept map below, ITS depicts all the characteristics of SoS, so intelligent transport system is definitely coming under the umbrella of the acknowledged SoS [3].

Each constituent system in ITS has a designated manager and resources for each SoS. However, the constituent systems return their independent ownership, objectives, funding, development, and sustainment approaches and hence ITS is categorized as acknowledged systems (Fig. 7–1).

7.4 Swarm intelligence with Internet of Things for transportation

The transportation industry has given substantial focus on swarm intelligence (SI) [5]. "SI is an innovative branch of meta-heuristics derived from imitating the behavioral pattern of natural insects" [5,6]. Studies done by Teodorović[6] and Zhanget al. [7] show that the challenges faced in intelligent transport system can be solved to an extend by heuristic algorithms, evolutionary algorithms, ant colony, agent-based model and genetic algorithms.

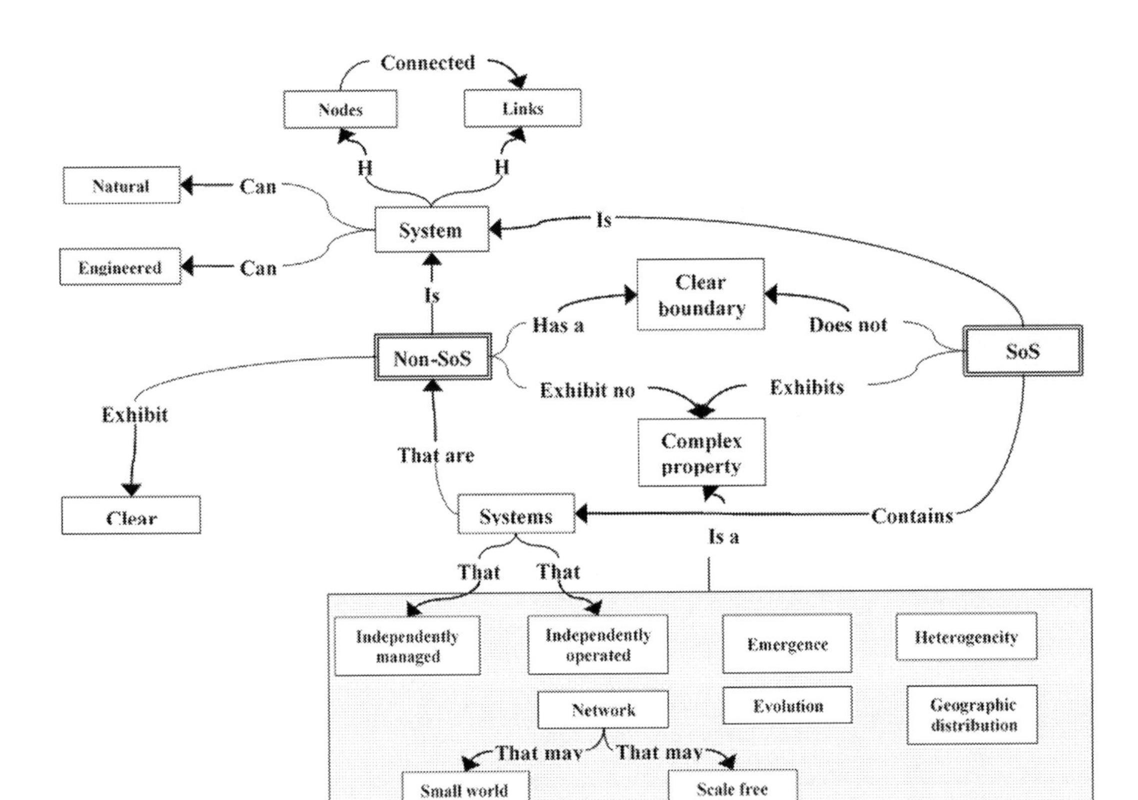

FIGURE 7–1 Concept map of SoS. *Adapted from M. Bjelkemyr, D. Semere, B. Lindberg, An Engineering Systems Perspective on System of Systems Methodology, 2009, DOI: 10.1109/SYSTEMS.2007.374659 [4].*

Moreover, a transportation system must be implemented, interrelated, and intelligent. In the process of finding a novel technology to support an intelligent transport system, the system should consist of "mobile communications, cloud technologies, energy storage, autonomous vehicles and the Internet of Things (IoT)" [8,9]. By uninterruptedly gathering, evaluating, and restructuring transportation information, IoT networks can render valuable, real-time information to both travelers and operators. Hence, they can support and improve the operations of ITS, traffic, and public transportation systems [5,10,11].

As per Marco Dorigo and Mauro Birattari [12], SI is the discipline that deals with natural and artificial systems composed of many individuals that coordinate using decentralized control and self-organization. SI can have more than one system and so is considered as having a SoS. In this chapter intelligent transport system is equated to SoS rather than restricting it to SI only.

7.5 Challenges in system of systems

1. Emergence—Unpredictable emergent behavior (and particularly undesirable emergent behavior) is the fundamental challenge of SoS. That is to say the development of theories,

methods, and tools to predict the emergent behaviors of combinations of SoS forms a critical underlying challenge for SoSE [8].

2. Effective systems architecting—The challenges are to develop a better understanding of the relationship of architectural features to behaviors, the assessment and evaluation of architectures, and the use of architectures by a range of SoS stakeholders. The improved understanding must be realised through new methods and approaches to architecting [3].
3. Situational awareness—The root cause of many SoS failures can be attributed to a lack of situational awareness of either human or nonhuman decision-making agents. In particular, the inability of an agent to foresee and analyze the behavior of future SoS states. This is different from "bad" emergence where adverse behavior arises due to all systems within the SoS behaving normally (sometimes called a normal error). Instead, a lack of situational awareness leads an actor in the SoS to behave inappropriately. This could be because of partial information, too much information, or misanalysis of information [3].
4. Governance and relationships—There are two main challenges associated with this theme: firstly, clarity of responsibility in SoS and, secondly, the approaches that are needed to encourage the owners/developers of individual systems to build and manage their systems in such a way that the overall goals of any SoS, within which such systems are incorporated, are achieved and/or the SoS benefits are maximized [3].
5. This requires research at quite a fundamental level into models of SA in SoS, information management, and presentation and also training and education for both systems developers and systems users [3].
6. The evolutionary nature of SoS causes issues with the interoperation of component systems that have been designed and built at different times, maybe to different standards, and with different performance characteristics [3].

7.6 Cyberphysical systems and limitations

These concern the increasing use of embedded software and, in particular, the direct and real-time interface between the virtual and physical worlds, including the IoT! There are many similarities with SoS, but the main distinction concerns emphasis. CPS is mainly concerned with control, whereas SoSE is mainly concerned with configurations, authorities and responsibilities, interoperability, and emergent behavior. CPS could be considered to be a special case of a SoS [8].

As demonstrated in Fig. 7–2, CPS is constantly trying to join with physical and cyber aspects of the concerned system, which might be in a network and the software in use will be in focus [13].

The main restrictions of CPSs, from a systems software perspective, are as follows:

- There is an incompatibility between application needs and system service supplies due to insufficient APIs making it difficult for applications to precisely identify the service they desire [8];

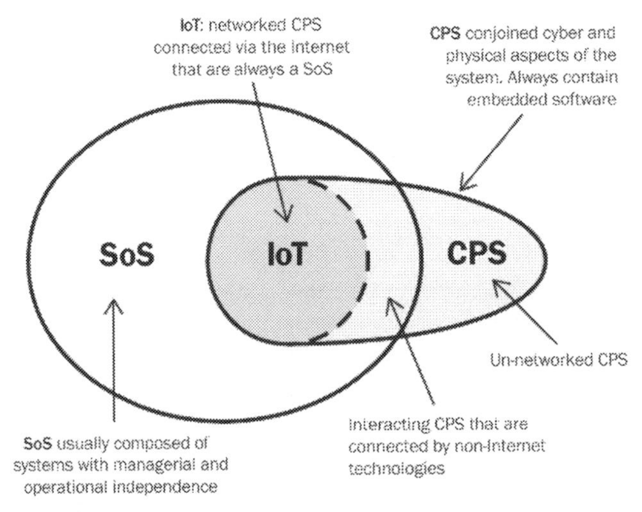

FIGURE 7–2 Relationship of SoS, CPS, IoT [13].

- System services tend to be application-agnostic, typically focusing on fairness and efficiency rather than timing, safety, security and reliability constraints; and
- Current systems are too inflexible to be easily extended with specific services for target applications [8].

The challenges faced by future systems support for CPS include:

- The design of an underlying software architecture that organizes itself with the most appropriate methods of communication and isolation between services [8];
- The automatic composition of services to satisfy application constraints, given underlying hardware limitations; and
- Careful design of APIs and consideration of hardware heterogeneity in the generation and verification of a software system for a given application [8].

In summary, the unresolved questions are:

1. How to guarantee the functional correctness of complex interactions between related vehicle features.
2. How to guarantee compliance with parafunctional requirements in distributed implementations of vehicle features.
3. What types of affordable run-time architectures to deploy to maintain prescribed levels of functionality and performance (ranging from full to degrade to failsafe) in the presence of hardware and software faults?

7.7 Problem statement

Due to the fast development of software engineering, it is important to identify the challenges and readiness of the emerging technology [8]. This chapter attempts to identify possible trends of software engineering technologies 30 years from now, and their impact on society and culture.

Every year many people grieve due to traffic casualities, which compels governments to enhance the degree of safety on roads by having a sustainable ecosystem [9]. Furthermore, passive safety systems, such as seat belts, airbags, and strong body structures, have not been able to lessen the number of accidents, so dynamic safety systems should be predominant. These include speed, direction, position, and acceleration information being tracked as a base of security measures. The concept of iM (intelligent mobility) is evolving as consumers, transport experts, industries, and governments realize opportunities and refine varied results by considering consumer-centric methodology for mobility opportunities as part of a wider, integrated system [14]. As per the www.smartmobilityuae.com survey, road deaths are given below.

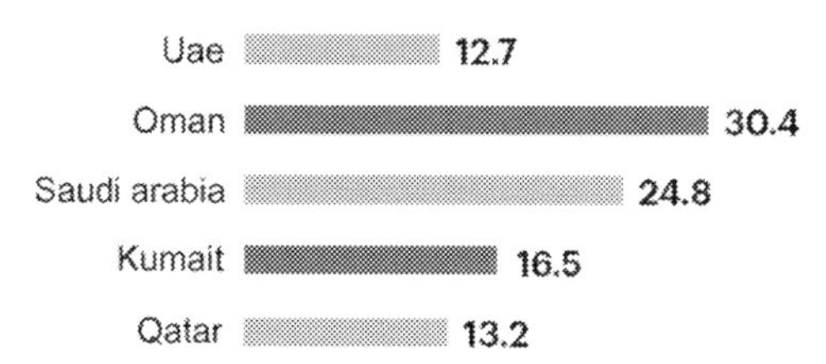

National highways and urban roads are drastically crowded and carbon emissions add to global warming, so the government is looking at a substitute for the current transport systems, which are sustainable, and to enhance safety and air quality via the use of new technologies like smart mobility.

7.8 Challenges in intelligent transportation systems

The challenges expected in technical and nontechnical areas are listed below:

1. AI—Empowering self-governing administration, better controlling of procedures and declines in cost [9].
2. IoT—Progressions in what it can process and the speed with superior networks. In transportation, we can solve challenges with rail safety, revenue collection, congestion management, and customer experience with intelligent IoT architectures and analytics software.
3. Energy—The concept of alternate energy resources instead of fossil fuels are a high priority and experiments are still ongoing.
4. 3D printing—Manufacturing is moving toward "weightless distribution."
5. Big data—Big Data, IoT, and AI together are considered the superhero team that is currently emerging.

However, there are still some obstacles in both the technical and nontechnical sides. Amongst the technical barriers are the present inadequate understanding of AI in the real world, the absence of standards and protocols for proficient operation of forthcoming transport networks, and the slow development in streamlining energy networks for transport and other needs [9].

At the same time, in nontechnical challenges, there should be a new guiding environment that is well-matched with autonomous vehicles and networks, safety and security, etc. Moreover, the communities should adjust themselves to the new disruptive changes to their daily life and accept the impacts on jobs, skills, and employment. Most importantly, coordination plans will be the expansion of all-encompassing information networks and big data applications that are safe and secure to enable the whole transport collaboration to operate successfully and proficiently and to establish flexibility and fast recovery when situations mandate this [9].

The need to develop autonomous vehicles in all transport modes as CPS devices that are harmless, protected, and competent to fulfill customer needs is a mandate from the perception of technology. Moreover, there is a vital requirement to provide for human authority and responsibility for their operations. It is also equally important to have standards and protocols for the interoperation of these vehicles. Simultaneously, the information networks will make available the on-the-spot information essential for smooth, safe operation. Everything necessitates support—the links, the vehicles, the information, and knowledge—both over the life cycles of the individual components and of the whole Systems of Systems represented beneath, as modification frequently happens [15]. This is a demand for systems for education.

7.9 Opportunities in intelligent transportation systems

Fully simulated autonomous vehicles still have a long way to go, but the opportunities to implement them are plentiful.

- Access to all—Even though the figures show there is a deterioration in the road accidents in the UAE, that drop is not a big one. Still the futurists of the country are trying to investigate ways to decrease road accidents, most of which is caused by inattentiveness, sloppiness, alcoholism, or wrong decision-making. A better safety and warning system that also has better security features with the latest sensing and control systems could make a big difference in the statistics. Additionally, trials are being carried in the area of lane changing, automatic braking systems, automated parallel parking, and autonomous override control [16].
- Changing markets—Since the sensors are becoming cheaper day by day, the vendors are utilizing them to the maximum, which will make the active control systems more reliable and verifiable [16].
- Demand for new infrastructure—The recent revolution in transportation like metro and transit has been overlooked on the arrival of accompanying services, such as car sharing or bike sharing [17]. Freedom in transportation is an ongoing challenge in the region. It is still a dream of both people in the city and villages to have a convenient pod which travels across places, communities, and sectors [18].

- Behavioral change—The competitors in the field could make a system that can give more customer satisfaction by enabling the vehicle facility with more flexible and faster service. The latest developments in big data analytics have helped the new entrants to understand the change in behavior of customer demands, allowing more flexible journey planning on demand and better monitoring services to vehicles [16].
- Personalized service—The demand of personalized mobility is increasing day by day giving more opportunity to enhance the standardization of transportation by allowing the customers to choose the seating and standing capacity on public transport [19]. This will allow flexibility to serve all users, including socially challenged and aging populations. The custom-made service which places an emphasis on customer satisfaction will help in breaching the obstacle of privacy. Moreover, the personalized services will prompt the customers to have more trust in users and the agents will get a chance to give better service to the customer [18].
- 5G network—It launches a wide-ranging wireless network with no limitation. Wireless World Wide Web, www is highly supported by the 5G network. The Telecommunications Regulatory Authority (TRA) has already announced the launch of the 5G network in Dubai, which will revolutionize the internet usage of UAE [20,21]. This technology falls in line with the directives of prudent leaders and UAE Vision 2021.
- Industrial revolution—The intelligent mobility has changed a lot due to the revolution in the industrial sector. Firstly mobility as a service has concentrated on customer-centric transportation to deliver an integrated transportation system [22]. Then big data has also traveled a long way, to where bid data analytics now has great influence on decision-making and prediction. As a result the smart roads are expected to give a maximum benefit and provide a better driving experience [23]. Recently IoT has become the trend revolution with the interconnected communication of vehicles, especially with emergency services [24]. Using a swarm system approach provides the agents with an ability to communicate both directly and indirectly [25].

7.10 Sustainability in intelligent transportation systems

1. Traffic management—ITS deployment is expected to increase the efficiency of traffic management. The biggest impact would be in reducing congestion and cutting down on travel time. By giving real-time traffic information to whoever is driving can enhance traffic efficiency. By reducing congestion and better traffic efficiency mobility becomes more punctual and reliable and thus increases the trust in customers. It also includes a quick response from transport managers to traffic incidents [15].
2. Carbon emission—Effective and efficient ITS will result in the public using more public transport which in turn will result in a reduction of carbon dioxide (CO_2) and other airborne toxins. Moreover, the vehicles integrated with ITS services can yield an approximate 15% reduction in carbon emissions. People's use of public transportation will likely increase to that of personally owned vehicles [15].

3. Economic values—The enhanced efficiency in ITS will yield economic dividends as well. This is achieved by optimized use of existing infrastructure and transportation systems. Simultaneously, moving both people and freight will benefit overall economic activity and traffic management. Moreover, it will assist in the development of industrial sectors such as automobiles, electrical equipment, communication networks, software, and engineering. Eventually, all of these will result in a reduction of time, costs, and the stress associated with travel time [15].

7.11 Intelligent transportation systems applications

The major applications used in ITS fall into five groups (Table 7–3).

ITS provides five key modules of benefits by (1) enhancing safety, (2) refining operational performance, especially by decreasing bottlenecks, (3) improving freedom of movement and accessibility, (4) supplying eco-friendly benefits, and (5) furthering throughput and growing economic and occupational growth [26,27].

7.12 Intelligent transportation systems products

1. Technology and Infrastructure Support—Advancement and installation of the dedicated short-range communication (DSRC), the global positioning system (GPS) and communication air interface and long and medium range (CALM) technologies improved the direction-finding and monitoring systems. Being a short-range communication protocol, DSRC offers a consistent and economical means for V2V and V2I interaction [14,28].
2. "Vehicle-to-Everything" Communication—The emerging core technology, V2X, transmits valuable traffic statistics with others using DSRC, GPS, IoT, big data analytics, etc. ITS countries are directing their research and business development into V2X, which is expected to be a doorway to the future of this communication technology [28].

Table 7–3 ITS applications.

Categories	Used for
Advanced traveler information systems	Providing drivers with real-time information, such as transit routes and schedules; navigation directions; and information about delays due to congestion, accidents, weather conditions, or road repair work
Advanced transportation management systems	Include traffic control devices, such as traffic signals, ramp meters, variable message signs (VMS), and traffic operations centers
ITS-enabled transportation pricing systems	Include systems such as electronic toll collection, congestion pricing, fee-based express lanes, and vehicle miles traveled usage-based fee systems
Advanced public transportation systems	Allow trains and buses to report their position so passengers can be informed of their real-time status (arrival and departure information)
Fully integrated intelligent transportation systems	Vehicle-to-infrastructure (V2I) and vehicle-to-vehicle (V2V) integration, enable communication among assets in the transportation system, for example, from vehicles to roadside sensors, traffic lights, and other vehicles

3. The Cooperative ITS Model—The advanced "ubiquitous" communication technology enhances the communication between systems and roads, vehicles and drivers by using C-ITS. The International Organization of Standardization (ISO) defines C-ITS stations facilitate actions that result in improved safety, sustainability, efficiency and comfort beyond the range of unconnected systems. These modern systems use a wireless connection to communicate with other vehicles or roadside infrastructure [14].
4. Advanced Transport Management Systems—Advanced Transport Management Systems operate with infrastructure fitted with vehicle detection systems, automatic vehicle identification, and CCTV to allow for real-time traffic data to be both sent and delivered through various service devices or facilities including VMS, the World Wide Web or mobile devices. The data collected are transformed into various ITS services including real-time traffic information, bus information services, and electronic toll collection services [14].
5. SoS Interference—The SoS field is also promising more advancement in prediction and trust systems by installing 18–20 different sensors in the vehicle. A few examples are light detection and ranging, radar, cameras, and GPS mapping. AI will provide advanced solutions for validation and verification, which will include how a smart car reacts to car crashes in the road. Vigorous deviations will also reach the network system as it is probable to have the 6G network via air instead of regular fiber optic and cables and broadband calculations [14].

7.13 An example: how can we make a change with intelligent mobility?

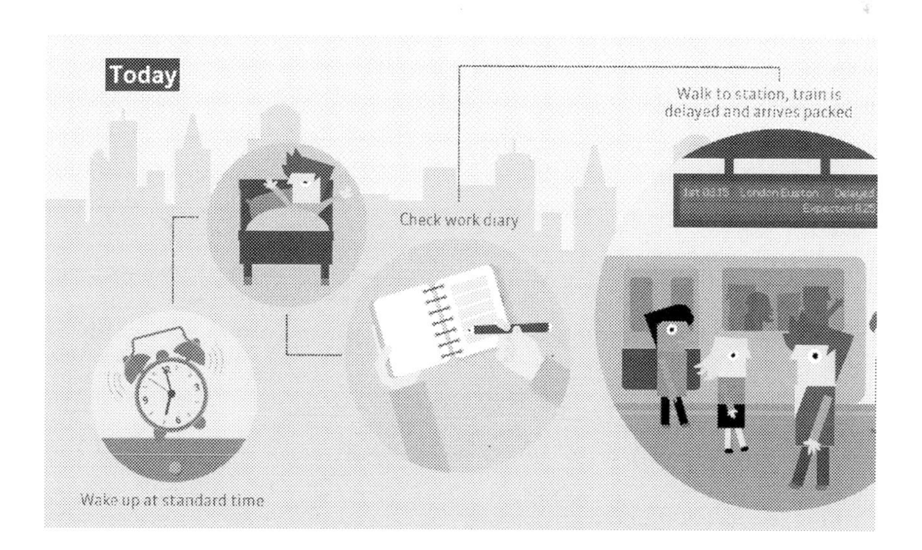

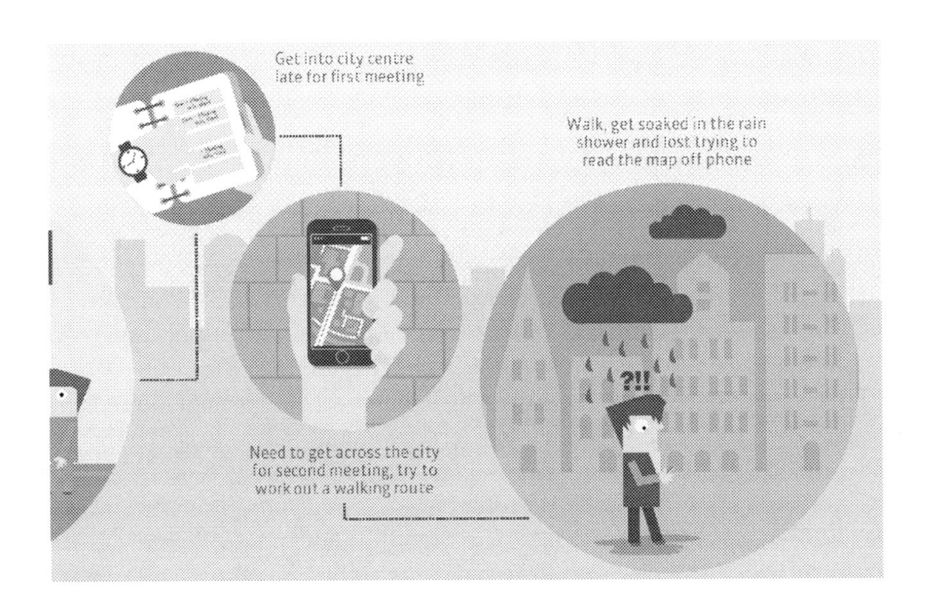
Get into city centre
late for first meeting
Walk, get soaked in the rain
shower and lost trying to
read the map off phone
Need to get across the city
for second meeting, try to
work out a walking route

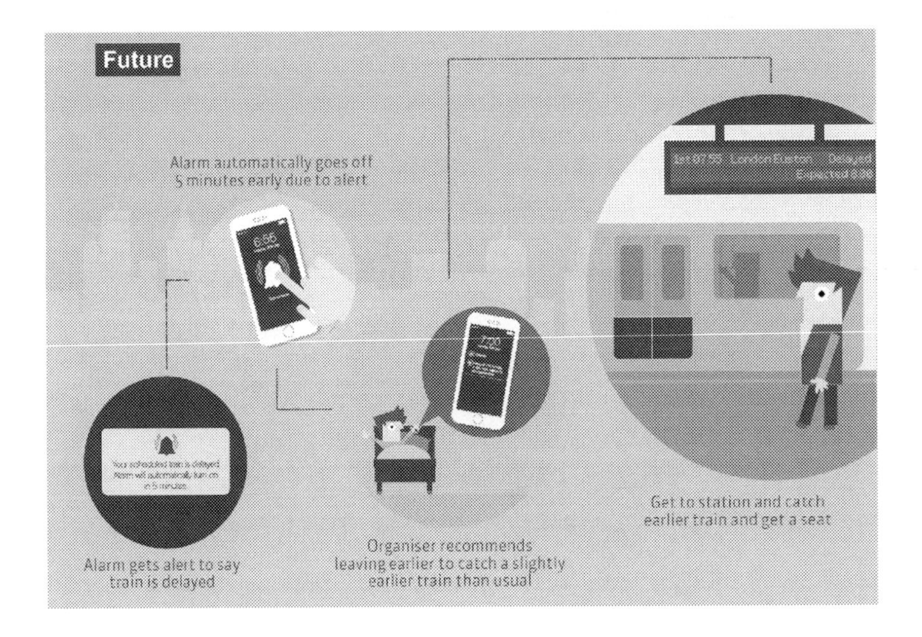
Future
Alarm automatically goes off
5 minutes early due to alert
Get to station and catch
earlier train and get a seat
Alarm gets alert to say
train is delayed
Organiser recommends
leaving earlier to catch a slightly
earlier train than usual

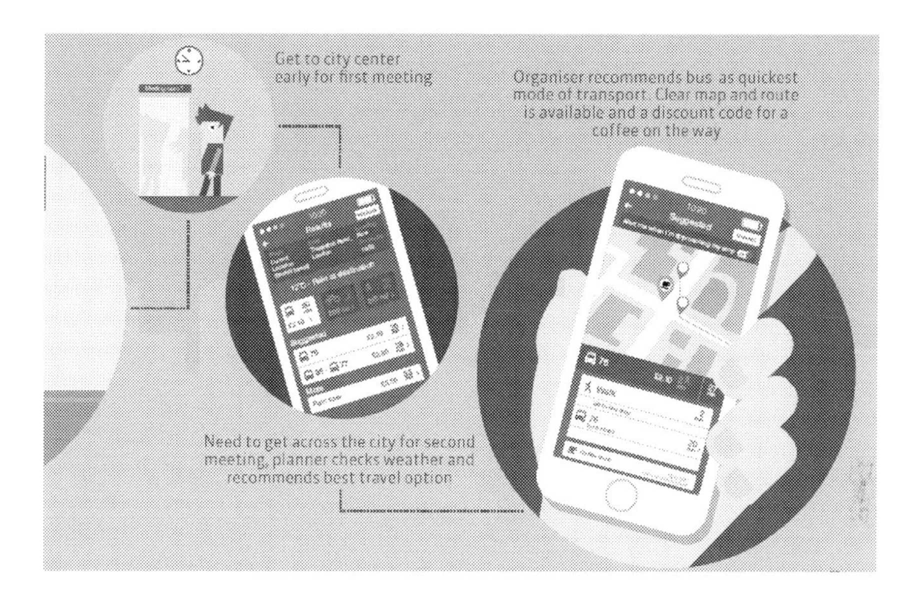

Adapted from Atkinsglobal.com. Available at: <https://www.atkinsglobal.com/~/media/Files/A/Atkins-Corporate/uk-and-europe/uk-thought-leadership/reports/Journeys%20of%20the%20Future_300315.pdf>, 2019 (accessed 24.09.19) [29].

7.14 The way forward

ITS needs more in-depth and systematic research in policy making. Autonomous vehicles will revolutionize the transport sector with numerous significant policy suggestions. Further research is required to measure how much ITS can lessen the bottlenecks and contamination and thus make a better sustainable ecology. Also, similar in-depth research is required to enhance safer and more effectual rail systems to improve the expedition efficacy and cross-border enablement [14]. The secretariat of the Economic and Social Commission for Asia and the Pacific could also bring out additional investigations into mixed mode commuting and interconnected transport systems, with a vision to promote smooth trade in possessions and the development of tourism in the province. A similar study should also be done in domestic waterways, navigation, air traffic control, and maritime transport [14].

References

[1] G. Xiong, F. Zhu, X. Liu, X. Dong, W. Huang, S. Chen, et al., Cyber-physical-social system in intelligent transportation, IEEE/CAA J. Automat. Sin. 2 (3) (2015) 320–333.

[2] A. Gokhale, M. McDonald, S. Drager, W. McKeever, A cyber-physical systems perspective on the real-time and reliable dissemination of information in intelligent transportation systems, Netw. Protoc. Algorithms 2 (3) (2010).

[3] K. Ravindran, Cyber-physical-social system in intelligent transportation, IEEE/CAA J. Automat. Sin. 2 (3) (2015) 320–333.

[4] M. Bjelkemyr, D. Semere, B. Lindberg, An Engineering Systems Perspective on System of Systems Methodology, 2009, https://doi.org/10.1109/SYSTEMS.2007.374659

[5] Z. Khan, A. Shawkat Ali, Z. Riaz, Computational intelligence for decision support in cyber-physical systems, 2017.

[6] D. Teodorović, Swarm intelligence systems for transportation engineering: principles and applications, Transp. Res. C: Emerg. Technol. 16 (6) (2008) 651–667.

[7] S. Zhang, C.K.M. Lee, H.K. Chan, K.L. Choy, Z. Wu, Swarm intelligence applied in green logistics: a literature review, Eng. Appl. Artif. Intell. 37 (2015) 154–169.

[8] A. Sadek, B. "Brian" Park, M. Cetin, Special issue on cyber transportation systems and connected vehicle research, J. Intell. Transp. Syst. 20 (1) (2014) 1–3.

[9] M. Elloumi, S. Kamoun, Adaptive control scheme for large-scale interconnected systems described by Hammerstein Models, Asian J. Control. 19 (3) (2017) 1075–1088.

[10] L. Atzori, A. Iera, F. Morabito, The internet of things: a survey, Comput. Netw. 54 (15) (2010) 2787–2805.

[11] P. Vlacheas, R. Giaffreda, V. Stavroulaki, D. Kelaidonis, V. Foteinos, G. Poulios, et al., Enabling smart cities through a cognitive management framework for the internet of things, IEEE Commun. Mag. 51 (6) (2013) 102–111.

[12] Marco Dorigo, Mauro Birattari, Swarm intelligence, Scholarpedia, 2(9) (2007) 1462.

[13] M. de C Henshaw, Systems of systems, cyber-physical systems, the Internet-of-Things. . .whatever next? Insight 19 (3) (2016) 51–54.

[14] The Information and Communications Technology and Disaster Risk Reduction Division, Intelligent Transportation Systems for Sustainable Development in Asia and the Pacific, ESCAP, UN, 2018.

[15] V. Kant, Cyber-physical systems as sociotechnical systems: a view towards human-technology interaction, Cyber-Phys. Syst. 2 (1–4) (2016) 75–109.

[16] Smart Mobility UAE Forum, Smart Mobility UAE Forum, 2018. Available: <https://smartmobilityuae.iqpc.ae/> (accessed 21.03.18).

[17] Google Drive - Cloud Storage & File Backup for Photos, Docs & More, Drive.google.com, 2018.

[18] E. Uhlemann, Transport ministers around the world support connected vehicles [connected vehicles], IEEE Vehi. Technol. Mag. 11 (2) (2016) 19–23.

[19] E. Mathew, S. Al Mansoori, Vision 2050 of the UAE in intelligent mobility, in: 2018 Fifth HCT Information Technology Trends (ITT), Dubai, United Arab Emirates, 2018, pp. 213–218.

[20] R. Team, A smart information system for public transportation using IoT, Int. J. Recent Trends Eng. Res. 3 (4) (2017) 222–230.

[21] A. Nuzzolo, A. Comi, Advanced public transport and intelligent transport systems: new modelling challenges, Transportmetrica A: Transp. Sci. 12 (8) (2016) 674–699.

[22] Robotics Editorial Office, et al. Acknowledgement to reviewers of robotics in 2016, Robotics, 6(4) (2017) 1.

[23] L. Pavlova, Wi-Fi and IoT in focus, LastMile 69 (8) (2017) 56–60.

[24] L. Tomory, Technology in the British industrial revolution, Hist. Compass 14 (4) (2016) 15.

[25] E. Issa Abdul Kareem, Traffic light controller module based on particle swarm optimization (PSO), Am. J. Artif. Intell. 2 (1) (2018) 7.

[26] Reimann, M., Rückriegel, C., Ingram, C. & Fitzgerald, J. (2017). Road2CPS Priorities and Recommendations for Research and Innovation in Cyber-Physical Systems. Karlsruhe, Germany: Steinbeis-Editions.

[27] M. Waldrop, Autonomous vehicles: no drivers required, Nature 518 (7537) (2015) 20–23.

[28] J. Mervis. Are we going too fast on driverless cars? Available at: < https://www.sciencemag.org/news/2017/12/are-we-going-too-fast-driverless-cars >, 2017 (accessed 23.06.20).

[29] Atkinsglobal.com. Available at: <https://www.atkinsglobal.com/~/media/Files/A/Atkins-Corporate/uk-and-europe/uk-thought-leadership/reports/Journeys%20of%20the%20Future_300315.pdf>, 2019 (accessed 24.09.19).

Further reading

M. Alam, Why the auto industry should embrace Blockchain. Connected Car Tech. Available at: <http://www.connectedcar-news.com/news/2016/dec/09/why-auto-industry-should-embrace-blockchain/>, 2019 (accessed 16.03.19).

M. Campbell, M. Egerstedt, J. How, R. Murray, Autonomous driving in urban environments: approaches, lessons and challenges, Philos. Trans. R. Soc. A: Math. Phys. Eng. Sci. 368 (1928) (2010) 4649–4672.

Guest Editorial, Driver distraction and inattention: meeting the challenges of new technology and automation, IET Intell.Trans. Syst. 12 (6) (2018) 405–406.

L. Misauer, IoT, Big Data and AI – the new 'Superpowers' in the digital universe | Striata, Striata. [Online]. Available at: <https://striata.com/posts/iot-big-data-and-ai-the-new-superpowers-in-the-digital-universe/>, 2018 (accessed 13.06.18).

Role of Internet of Things and image processing for the development of agriculture robots

Parminder Singh[1], Avinash Kaur[1], Anand Nayyar[2]

[1]SCHOOL OF COMPUTER SCIENCE AND ENGINEERING, LOVELY PROFESSIONAL UNIVERSITY, PHAGWARA, INDIA [2]DUY TAN UNIVERSITY, DA NANG, VIET NAM

8.1 Introduction

The industry of robotics was originally developed to replace or supplement humans by doing repetitive, dull, dangerous, or dirty work [1]. Robot systems have various applications in agriculture, industry, defense, and other fields. Various research is carried by the application of agriculture robots and the automation of various greenhouse and field operations. The feasibility and technical fundamentals are widely demonstrated. Adverse interference, an unstructured environment, and diversified and complicated operation process are the main things obstructing the commercialization of robotics in agriculture operations. Unlike the manufacturing industry, robotics in agriculture does not share the same success rate [2]. The reason is due to the less structured agriculture environment and flexibility of varying agriculture objects [3]. Also, it is difficult to make robots adaptive for the automation of the agriculture process.

In recent years, different kinds of agriculture robots have achieved success due to the development of smart sensors, agriculture techniques, and information techniques. The main areas of research in agriculture robots are robotic grasping, localization and mapping, object recognition and location, and collaboration operation.

8.1.1 Recent trends in agriculture robots

Agriculture Environments Simultaneous Localization and Mapping (SLAM): It is one of the main problems in automatic robotic navigation and positioning. The external information is extracted using multiple sensors to obtain a consistent environment map and also obtain recognition of itself within this map [4]. The SLAM problem solutions are divided into three main categories: light detection and ranging (LiDAR), visual, and sensor fusion. The different kinds of SLAM algorithms are an indoor autonomous mobile robot, virtual reality or

Swarm Intelligence for Resource Management in Internet of Things. DOI: https://doi.org/10.1016/B978-0-12-818287-1.00014-0

augmented reality equipment, and unmanned vehicle [5]. The SLAM is applied to agriculture with the development of SLAM technology. It is a mixture of new technologies and traditional fields. The main goal of precision in agriculture is to locate a moving vehicle accurately [6]. By finding the solution for SLAM, mapping of the target area can be performed by the vehicle and it can locate itself and also perform other tasks of mowing, weeding, and spraying [6,7]. The farmers' workload can be reduced by using the SLAM technology. The dangerous tasks can be completed replacing humans and productivity can be increased [8]. SLAM technology has bright future prospects for the intelligence and automation of agriculture.

Harvesting robots object recognition: Recently, the use of robotic systems in agriculture has been trending [9]. The precondition of robot grasping is the recognition of an object. The method utilized by the harvesting robot is based on the vision of a computer for achieving the location and recognition of objects. The processing procedures are the key step of computer vision, processing visual images acquired by the visual sensors including the object. The acquired images consist of much irrelevant information to the object. This reduces the accuracy and speed of recognition. Hence, extracting useful information while detecting vegetables or fruits is an important part.

The researchers introduced feature engineering [10]. The feature is the available information in recognition of the object. The difference between the growing environment of the object and the object can be figured by using and extracting those features. But there are some complex and uncontrollable factors near vegetables and fruits. The difficultly of location and recognition is enhanced by illumination unevenness on the surface of the fruit, occlusion of stems and leaves, and high variation in fruit color. Hence, there arises a need to develop a robust algorithm to be applied to the recognition of objects for harvesting robots.

Product quality sensing in agriculture for grading robots: In daily life, vegetables and fruits have become indispensable food. The vegetables and fruits are not only important for taste but also contain trace elements, fiber, and vitamins. Recently, the external and internal quality of agricultural products has become of great significance for the people [11]. The automatic grading of agricultural products is playing an important role in the field of agriculture due to the increasing demand for the safety and security of food. The need for human operators to perform dangerous, monotonous, and heavy operations is replaced by automated robots. Hence, in small-value farming operations, it can increase economic sustainability. The main goal of an automatic grading robot is to detect external defects in vegetables and fruits using machine vision. It also measures the internal quality by using spectral/hyperspectral imaging technology [12]. Hence, the processes for grading vegetables and fruits are becoming highly automated robotics technologies and machine vision.

Robotic grippers and grasping control: The key tasks for robotic manipulators are the holding and grasping of objects. The grippers are the mechanical interface between robots and the environment of robots. They are the most critical components for performing the manipulation of tasks. The vegetables and fruits are susceptible to bruising due to the early performance of harvest and postharvest processes. The damage caused due to the automated grasping process is the key barrier to the replacement of manual labor.

The tasks of agriculture are performed in an outdoor environment [13]. The sequence of operations and agricultural tasks vary from task to task. The grippers and manipulating objects of robots are flexible and vary in structures, sizes, and shapes. Hence, there arises the need for grasping control strategy, smart sensors, and robotic grippers in agricultural tasks. The robotic grippers are designed with an aim to grasp any kind of object by copying human abilities, such as visual perception and sense of touch [1]. The robotic grippers integrate various sensors not only for manipulating the workpiece but also for analysis. Online decision-making can be performed based on fusion sensory data.

8.2 Research in agriculture robotics

The wide range of applications is covered by works of research on agriculture robotics, such as the use of autonomous target spraying for pest control [14]. Field robots of agriculture contribute to improved yield, improved soil health, and increased reliability of operations [15]. They are equipped with multiple cameras and sensors for path planning algorithms, localization and mapping, and navigation control [16–18]. A recent achievement in agricultural robotics is the harvesting platform SWEEPER EU H2020 in July 2018. It possesses a catching device for the harvesting of fruits. It detects mature fruits. The setup of the sensor and camera is independent of conditions of light. It also detects obstacles such as plant stems and leaves. The recent advances in different fields are discussed in the following subsections.

8.2.1 Weed control and targeted spraying robot

The main aim of agriculture robotics is the replacement of the workforce of humans by field robots for handling the tasks with more efficiency and accuracy [19–21]. The most demanded applications are precise spraying and weed control. Targeted spraying reduces the herbicide usage in the application of weed control [22]. Fig. 8–1 shows the different weed control robots available: ecoRobotix (Yverdon-les-Bains, Switzerland), multipurpose farming and weeding robot platform named BoniRob [23,24]; AgBot II [25], an innovative field robot prototype created by the Queensland University of Technology for independent implementation of fertilizers, weed detection, and classification and mechanical or chemical control of weeds; autonomous robot [26], a weed control study robot created by the University of Applied Sciences in Osnabriick; Tertill [27], a fully independent compact solar-powered robot created for weed cutting by Franklin Robotics; Hortibot [28], a robot created by the Faculty of Agricultural Sciences at Aarhus University to transport and attach a range of weed detection and control instruments such as cameras, herbicides, and spray booms; and Kongskilde Robotti [29], a robotic platform fitted with a FroboMind software-based drive belt [30] that can be attached to various modules and tools for automated and semiautomated mechanical weed control, precision seeding, opening and cleaning of furrows.

FIGURE 8–1 Examples of weed control and target spraying robots. (A) BoniRob; (B) AgBot II; (C) Autonome Roboter; (D) Tertill; (E) Hortibot; (F) Kongskilde Robotti; (G) RIPPA; (H) Spray robot. *From (A) Deepfield Robotics; (B) Queensland University of Technology; (C) Osnabrück University; (D) franklinrobotics.com; (E) technologyreview.com; (F) conplecks.com; (G) The University of Sydney; (H) Hollandgreenmachine.*

8.2.2 Field scouting and data collection robots

Field scouting robots face different interdisciplinary difficulties in offering accurate information and measurements that precision farming and crop models can use and process. Apart from the problems of intrinsic physical and biological variation in farm areas and orchards, it is anticipated that scouting robot platforms will be versatile, multipurpose and inexpensive to be considered viable.

These robots can play a main role in lowering manufacturing costs, improving productivity and quality, and allowing tailored plant and crop treatments, if effectively incorporated and applied. Development of scouting robots for information collection and modern farming involves the comprehensive use of sophisticated sensors for precision farming [31,32] to produce useful outcomes while performing automatic and precise navigation control, manipulator control, barrier prevention, and three-dimensional reconstruction of the surroundings.

For instance, there was a proposal for an independent field survey mobile robot platform with a custom manipulator and gripper [33] to perform image sensors and GPS designs for independent navigation and information collection within greenhouses and open field cultivation settings, as shown in Fig. 8–2. Fig. 8–3 shows examples of general purpose robots for field scouting and data collection.

8.2.3 Harvesting robots

Traditional fresh market harvesting of fruits and vegetables is a labor-intensive job requiring a shift from tedious manual operation to continually automated harvesting. Increasing effectiveness and lowering harvesting-dependent labor will guarantee the output and competitiveness of high-tech food production. Despite developments in agricultural robotics, in open fields and greenhouses, millions of tons of fruits and vegetables are still hand-picked each year.

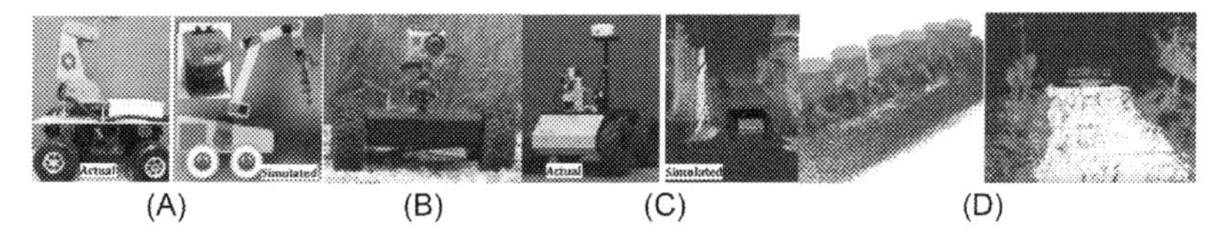

FIGURE 8–2 Examples of professional field robots for 3D reconstruction and scanning of plants. (A) A prototype surveillance field robot (AdaptiveAgroTech.com); (B) OSCAR field survey robot (Inspectorbots.com); (C) Husky UGV for field scouting and 3D mapping (Clearpathrobotics.com); (D) point cloud and detected maize plants, and 3D point clouds of vineyard created by VinBotRobotnik Automation (www.robotnik.eu).

FIGURE 8–3 Examples of general robots for data collection and field scouting. (A) TrimBot (trimbot2020.org); (B) Wall-Ye vineyard robot (wall-ye.com); (C) Lady bird (University of Sydney); (D) MARS (echord.eu/mars); (E) SMP S4 (smprobotics.com); (F) VineRobot (vinerobot.eu); (G) HV-100 Nursery Bot (harvestai.com); (H) VinBot [59] (vinbot.eu/); (I) Mantis [32] (University of Sydney); (J) GRAPE (grape-project.eu).

Apart from the elevated labor costs, the availability of qualified workers that accept repeated duties in the difficult circumstances of the field imposes uncertainty and cost. In order to be cost-effective, robotic harvesting requires maximization of the fruit yield to compensate for the extra cost of automation. This leads to crops growing at greater densities, making it even more difficult for an autonomous robot to detect, locate, and harvest the fruit at the same time.

In the case of sweet pepper fruit, with an estimated output of 1.9 million tons/year in Europe, studies show that while an average time of 6 seconds per fruit is needed for automated harvesting, the accessible technology has only reached a success rate of 33% with an average picking moment of 94 seconds per fruit [34].

R&D in robotic harvesting dates back to the 1980s, with Japan, the Netherlands, and the United States being the pioneering nations. The first trials used easy monochrome cameras inside the canopy to detect fruit [35]. Other than visible light RGB cameras [36] and ultrasonic radar sensors that are commonly used for object detection because of their affordable cost [37], advances in sensing and imaging technology have resulted in the use of

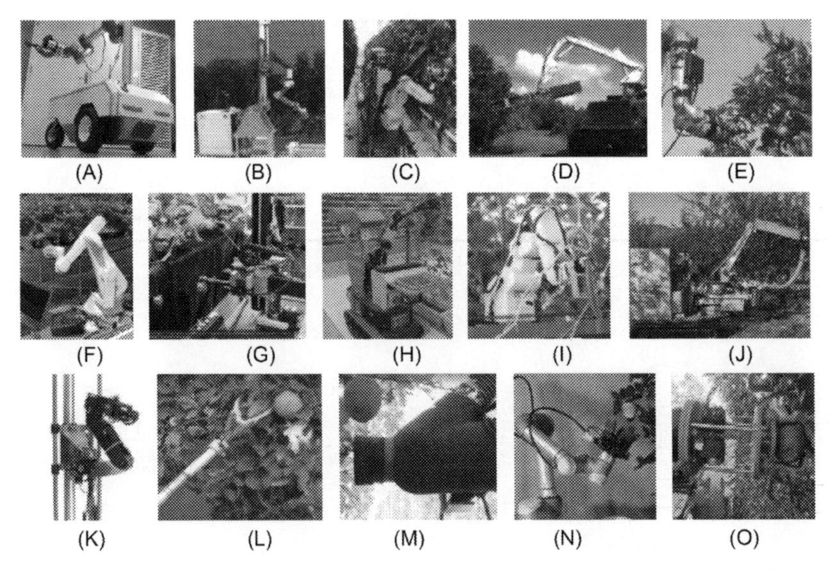

FIGURE 8–4 Examples of harvesting fruit robots. (A) Harvey (Queensland Univ. of Technology); (B) CROPS (crops-robots.eu); (C) SWEEPER (sweeper-robot.eu); (D) Energid citrus picking system (Energid technologies); (E) Citrus robot (University of Florida); (F) DogTooth (www.dogtooth.tech); (G) Shibuya Seiki (shibuya-sss.co.jp); (H) Tomoto harvesting robot (szbotian.com.cn); (I) cucumber robot (Wageningen UR); (J) Apple harvesting robot; (K) Apple harvesting (crops-robots.eu); (L) Apple picker (FFRobotics.com); (M) Apple picking vacuum (abundantrobotics.com); (N) UR5 apple robot (University of Sydney); (O) Apple catching (Washington State University).

sophisticated devices such as infrared [38], thermal [39], hyperspectral cameras [40], LiDAR [41], or multisensor combinations [42]. Examples of harvesting robots are shown in Fig. 8–4.

8.3 Digital farming in agriculture robots

Agricultural robotics is a promising solution for digital farming and managing labor shortage issues and decreasing profitability. Initial experiments with one of the latest techniques available for automated harvesting (the Harvey [28] robot) have already shown a success rate of 65% and a detachment rate of 90% for sweet pepper cultivation in actual planting scenarios where no leaves and occluded fruits have been trimmed or removed.

Field agent robots that track and collect data autonomously empower growers with comprehensive information on their plants and farms in real time, revealing upstream pictures to make data-driven choices. Agricultural robotics is taking farming methods into a fresh stage by becoming more intelligent, detecting sources of field variation, consuming less energy, and adjusting performance to more flexible duties. In the future manufacturing of vegetables and crops, that is, growing plants in space or the creation of robotized plant facilities in Antarctica, they will become an essential component of the larger picture.

The trend in food production is toward automated farming methods, compact agricubes, and crop systems with the minimum human interface where skilled labor is substituted by robotic arms and mobile platforms. In this context, digital agriculture has incorporated fresh ideas and sophisticated techniques into a single structure to provide farmers and stakeholders with a quick and reliable technique of plant-level real-time observation (i.e., field information collection and crop tracking) and more accurate action (i.e., diagnostics, strategic decision-making, and implementation).

The purpose of digital farming is to collect high-resolution field and weather data using ground-based or aerial-based sensors, transmit these data to a central advisory unit, interpret and extract information, and provide farmers, field robots, or agroindustries with choices and actions. Examples include a thermal-RGB imaging system for monitoring plants and soil for health assessment, creation of information maps (i.e., yield and density maps), and data sharing. Implementation of digital farming methods results in viable, effective, and stable manufacturing with substantial yield increases.

In a sustainable way, more and high-quality yields are expected from modern farms that are not dependent on labor. Digital farming is one of the solutions to this problem. It is dependent on the field data collection from agriculture robots which involves sensor technology. Agricultural producers, scientists, and farmers face the challenge of generating more food for meeting the demand of the world population of 9.8 billion by 2050 [43]. The integration of control technologies, sensors, and digital tools has put a spurt into the development and design of agriculture robots, hence, demonstrating benefits in modern farming. The evolution is being achieved by collecting detailed and accurate spatial and temporal data for the completion of nonlinear control tasks for robotic navigation. Automatic farm machinery and guided tractors embedded with global and local sensors for working in orchards and raw crops are mature now. An important aspect of digital farming is manipulators and agriculture field robots [44,45]. The application of robot technology in digital farming is proceeding toward automation. This is leading to a change from the traditional activities of the field to high-technology tasks of industries and is attracting companies, professional engineers, and investors. These robots perform many kinds of operations, such as milking [46,47], target spraying [48,49], sorting [50], phenotyping [24,51], and harvesting [13,52]. Such types of applications are difficult to automate. An agricultural robot can manipulate, sense, or touch the crop and its surroundings, leading to an increase in efficiency with minimal impact [53]. The uncertain tasks and unstructured environments impose great challenges, thus limiting the application of agriculture robots.

Robots with precision, speed, and accuracy are available but possess limited application in agriculture due to uncertain tasks and unstructured environments. As an example, the off-season cultivation of vegetables and fruits requires different kinds of robotics and automation in closed field plant production such as greenhouses [54]. A robot for the field in a dynamic environment with harvesting manipulator, deleafing, end-effector, and spraying should consider the factors of shapes and sizes of plants, branches, stems, leaves, weather influences, and texture. For example, in the case of harvesting the mechanism of sensing identifies the ripeness of fruits in an unpredictable heterogeneous environment. Hence, this is a more challenging environment than that for industrial robots.

8.4 Navigation algorithms in agriculture robotics

In this section, based on the measurements from a LiDAR detector, three distinct infield navigation algorithms are presented. The algorithms are evaluated on a tiny field robot. They are used between the two adjoining rows of labyrinth crops to drive the robot autonomously. In the first algorithm measurements from the right and left side take distance. The position of the robot is adjusted accordingly if it is not placed in the middle of the mid-row space. Another technique uses the least squares fit method for grouping the right and left measurements into two vertical rows. The course of the robot is adjusted according to the orientation of both lines and calculated range. The third technique tries to fit an optimal triangle between the plants and the robot. On the basis of shape, the course of the robot is adjusted.

8.4.1 Introduction

The technique for autonomous agriculture driving machinery in the field is to use a precise differential [55] or real-time kinematic global positioning systems [56]. For complete working, the previously known path of movement is required. This process is repeated with each iteration [57]. There may be a condition of unavailability of GPS data, so it makes it infeasible sometimes. Hence, under these conditions, distinct systems are to be employed. A most feasible solution is to use stereo cameras to construct a 3D point cloud [58] or to use cameras to detect plant lines [59]. Another possible method is the usage of LiDAR systems [60] which take specific measurements by reading distances from the sensor to the first obstacle and repeating this for the entire range (usually 270 or 360 degrees in steps of 1, 0.5, or even 2.25 degrees).

The machines [61] can move with the assistance of these technologies even in an unfamiliar sector where they were not previously implemented. They depend on the farm estate and the planted plants. The crops are parallel and the machines can drive in mid-row spaces between the plant lines so as not to harm the crops. Various approaches/algorithms can be used to direct the machines. More advanced systems even create an environment map using distinct SLAM methods [62,63] and locate the machines in a fresh environment that is only being found. Using a route planner [64] and path follower [65] to follow the route, it then builds a route based on the map. These methods are used in unfamiliar scenes but are not really essential in infield circumstances where previous plant pattern data is known. The aim of this job is, therefore, to first introduce three distinct algorithms that could be used to drive the machines autonomously and to assess their precision for navigation in the infield. All three were implemented as part of the robotic operating system [66] and applied on a small field robot, making it autonomous when driving through the field in mid-row spaces. Their purpose is not to plan and follow the path but to adjust the heading of the machines/robot at every measured location.

The different navigation algorithms are:

- *Minimal row offset-based algorithm*: The first algorithm is the easiest. As shown in Fig. 8–5 it requires 30 measurements from the left and 30 from the right. From these two

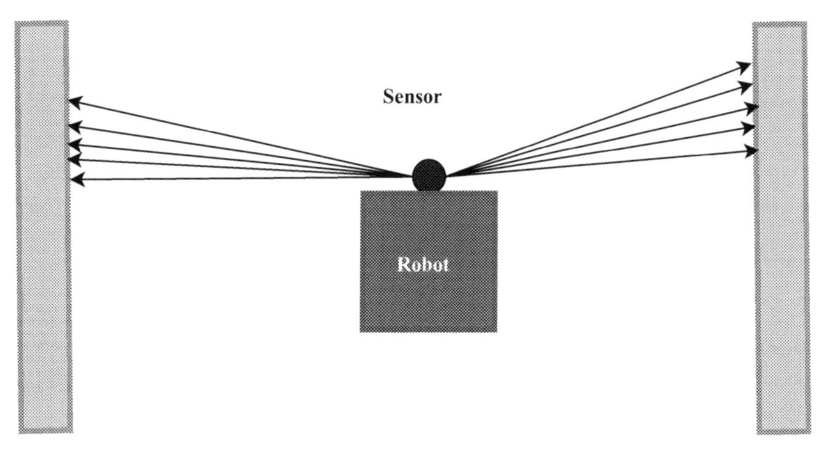

FIGURE 8–5 The robot is standing in the middle robot. The measurement of the sensor is represented by a double-sided arrow.

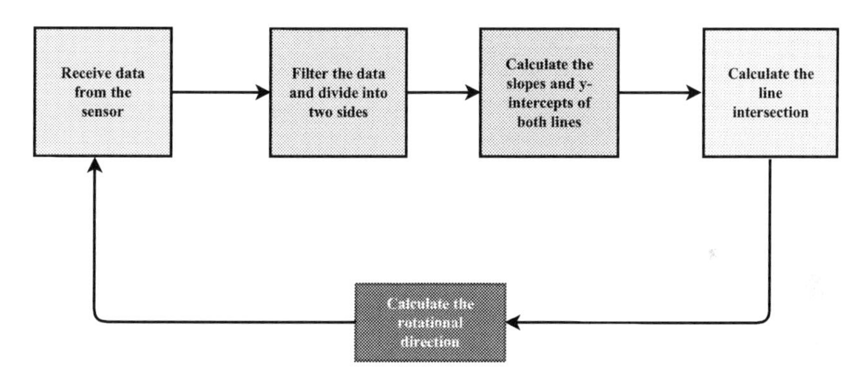

FIGURE 8–6 Each step in a cycle.

sets, it first eliminates ones that are too far away, that is more than 0.75 cm and belong to the other crop lines, and then calculates an average distance value for each side. These average distances can be written as h_l and h_r for the left and right sides, respectively. To calculate the offset these are then used as shown in Eq. (8.1).

$$\text{offset} = (h_r - h_l) \tag{8.1}$$

The course of the object is adjusted by using the value of the offset as in Eq. (8.2).

$$\text{Orientation} = \text{Offset}(h_r + h_l) \tag{8.2}$$

- *Least-squares fit approach algorithm*: This second algorithm was intended to navigate the robot between two parallel walls, either an artificial barrier, such as walls, or actual crop lines, such as labyrinthine crops. Fig. 8–6 shows the summary of the method, explaining each sensing–adjustment cycle. The detector reads the data and distances for each degree between the sensor and the barriers and triggers a callback feature every time the measuring sequence is completed.

The callback function first filters the data depending on the distance readings. The points that are too far away and the points that are too near to count are discarded. The algorithm makes it possible to set how many points should be included for each count for each side, corresponding to how many degrees will be used in the subsequent steps of the algorithm. Two information sets store the helpful measurements, one for the left and one for the right.

A linear fit for which a minimum square method [67] has been selected is used in the third phase. This calculates a linear equation's path and the y-intercept describing each set for each side. The least-squares method enables us to fit a lower amount of heavy-duty mathematical operations linearly to the readings. To do this, we need to identify some extra parameters to calculate each line's slope and y-intercept.

Once the information about the intersection from the two lines is known, the position of the robot is calculated. The distance stays constant if the robot is aligned up with the row, no matter which wall is closer. The L describes how far away it is from both walls/lines. With just looking at the two distances, the problem is simplified.

The strategy outlined can portray two distinct scenarios based on the robot's situation. The first scenario is when the robot is in the correct place and the second is when the robot is not in the correct place.

In both cases, the parameters of the two linear equations and the distance to the path of the robot are used. Since the walls are always apart from each other with a constant width, the triangle covers the same surface. With the robot, which generates distinct triangles, what changes are the orientation and position? If the robot is not aligned in parallel to the walls or crop lines, an asymmetric triangle is constructed.

- *Triangle-based navigation algorithm*: The third approach [68] to finding an optimal path for the robot consists of multiple steps. The algorithm uses trigonometric functions to calculate the distance of a segment between every two sequential points from the LiDAR sensor. The segment distance must be wide enough to drive the robot through, and if they do not meet the criteria, they are disposed of.

The algorithm calculates the angles: α, *beta*, and γ in a triangle limited between two sequential points and the LiDAR sensor. If any of the angles α or β is bigger than a predefined threshold, set to 100 degrees or more, which would produce a triangle that would not fit in field situations, the segment is disposed.

The algorithm calculates the angle for the front wheels to turn when determining the ideal section. The ideal segment midpoint is calculated to drive to the robot.

8.5 Intelligent detection of insects

Identification of corn pest illnesses is one of the farmers' main duties for addressing social and environmental problems, such as decreasing pollution induced by pesticide use and keeping grain production stability. The most prevalent maize pests are pyralide insects [69] and they do a good deal of damage to maize yield and quality.

Traditional manual surveillance needs not only a great deal of work but also creates detection due to human omissions. With the rapid growth of computer technology, computer vision-based surveillance of illnesses and insect pests was possible, which can significantly enhance the detection and recognition of pests in real time [70]. There are presently some methods available to detect plant or insect diseases [71] with image processing and computer vision technologies. For example, Ali et al. used color histograms and textural descriptors to recognize citrus illnesses [72].

They used the color difference to differentiate between the affected area and the disease. Lu et al. utilized spectroscopy technology to define crown rot in [73] strawberry anthracnose. Xie et al. used hyperspectral images to define the presence of gray mold disease in tomato leaves [74].

In addition, researchers have developed an automated detection and monitoring scheme to detect tiny pests in the greenhouse, such as whitefly, etc., which can effectively monitor tiny insects and their densities [75,76]. In the meantime, computer vision technology has also been used to detect and monitor their [77] populations. For parasites on strawberry crops, in the greenhouse setting [78], the support vector machine (SVM) method coupled with the picture handling technique was effective in identifying the thrips with an error of less than 2.5%.

To segment pests or any object from the [79] image, the incorporation of k-means clustering with image processing methodology was used. Dai and Man used a convolutionary Riemannian texture with differential entropic active contours to distinguish background areas and expose [80] plague regions.

Zhao et al. obtained an accurate contour of insect pests and plant diseases for the recognition, using the difference in active contour and texture variations [81]. In their further studies [82], they also introduced a method of image segmentation for fruits with diseases based on Otsu constraint and level set active contour. As far as insect and disease recognition is concerned, some latest developments in studies can be categorized into the two classifications. Image processing and computer vision techniques are the focus of the first category without the need for information coaching. It is a method of pesticide recognition based on multifunctional fusion and sparse depiction for defining beetles [83].

Four methods for the classification and diagnosis of the corn leaf diseases are presented by using machine vision techniques and image processing [84]. Martin et al. introduced an extended region growing algorithm, to identify and count the pest to predict the pesticide amount to be used [85]. Przybylowicz et al. developed a technique based on wing measurements, which can be an effective tool for monitoring the European corn borer [86].

The second category concentrated on the training of data models, which mainly used machine learning and neural network technology. The method based on the difference of the Gaussian filter and local configuration pattern algorithm was used to extract the invariant features of the pest images, and then these features were put to a linear SVM for pest recognition with a recognition rate of 89% [87].

Kohonen's self-organizing maps neural network was used to identify the extracted insect pests caught by a sticky trap [88]. In addition, Boniecki et al. proposed a classification neural model using optimized learning sets acquired based on the information encoded, which can be used to accurately identify the six most common apple pests [89].

Based on the mixture of an image processing algorithm and artificial neural networks (ANNs), Espinoza et al. suggested an algorithm for the detection and monitoring of adult whitefly (*Bemisia tabaci*) and thrip (*Frankliniella occidentalis*) in greenhouses, with the right identification speed above 0.92 [90]. In order to recognize insects, Zhu et al. combined the color histogram with the dualtree complex wavelet transform [91] and SVM [92], which can enhance insect recognition. Li et al. suggested a technique of red spider recognition based on the clustering of k-means that converted the picture into a clustering Lab color space [93]. To recognize the red spider with apparent red characteristics, this technique had a high precision rate. However, the technique can only be implemented if the color contrast between the objects and the scenes is large.

Furthermore, a picture acquisition device is also required [75]. Johannes et al. automatically provided a wheat disease diagnostic system through the use of portable capture devices [94]. In his studies, in conjunction with statistical inference techniques, a novel image processing algorithm based on candidate hotspot detection is suggested to address disease identification in wild circumstances.

The image processing and computer vision technology has been commonly used in the identification and recognition of illnesses and pests since the literature analysis in recent years and has accomplished excellent outcomes. The scientists generally used a current technique in conjunction with image processing methods to detect and recognize clustering methods, neural networks, texture analysis, wavelet transformation, level set technique, etc. However, having a universal technique for detecting and identifying all pests is hard. The algorithms are generally used to detect and recognize a pest or a pest class. Moreover, most of the current research is often directed at the greenhouse setting, and a practical verification scheme is generally not built by scientists. Obviously, high-precision identification can be achieved by deep learning, but this training-based strategy is hard to ensure in real time and needs a large amount of current information to train the model.

Relatively few studies on the detection and identification of insects from Pyralidae exist at present. We have studied the following elements to automatically and precisely detect and recognize the Pyralidae insects in real time. First, a pest surveillance robot platform is designed and manufactured. Then a Pyralidae insect identification system is provided, in which the image's color characteristic is used. The inverse mapping technique of the histogram and the multitemplate picture are used to achieve the overall superposition of the probability picture. Next, the Otsu constraint segments the picture. Finally, contours and Hu moments are used to automatically screen and recognize contours; thus it is possible to recognize the contour of Pyralidae insects.

Some techniques based on computer vision were suggested to detect pests/diseases of plants. They can generally be classified as uncontrolled and controlled techniques. The scientists need to evaluate the particular characteristics of pests/diseases in the category of unsupervised techniques and design a function extractor to separate pests/diseases from the background, then extract extra characteristics linked to the segmented region for use in the classification of pests/diseases.

Xia et al. applied multifractal analysis [75] to segment action of whitefly pictures based on local singularity and worldwide picture features. Dai and Mann [80] studied the Riemannian convolutionary texture structure to distinguish the environmental background textures and potential pest textures in pest camouflage in grains, and then a differential entropic active contour model was created to detect pests.

Xia et al. [95] used the watershed algorithm to segment pests from the background pictures to recognize prevalent plant pests, such as whiteflies, aphids, and thrips, and then pest color characteristics were extracted by Mahalanobis range to recognize pest species. The multifunctional fusion-based technique was also used to detect pests/diseases in addition to a single cue of function channels. The writers, for instance, use color, shape, and texture characteristics in [83] and combine them with sparse representation to identify pests.

As the evolution of machine learning technology has increased, more and more supervised models have been designed to define pests/illnesses. The characteristics of pests/diseases are described by humans in traditionally monitored teaching. The feature information is first computed in practical apps and then fed to appropriate classifiers to identify pests/diseases. For example, Ali et al. applied the color difference algorithm to separate the affected area from the disease and also used the color histogram and textural characteristics to classify diseases [72]. Lu et al. used spectroscopy technology to identify crown rot anthracnose in strawberry [73], using spectral vegetation indices to train stepwise discriminant analysis (SDA), Fisher discriminant analysis (FDA) and k-nearest neighbor (KNN) classifiers. Xie et al. used the KNN classifier to determine whether hyperspectral pictures show gray mold illness in tomato leaves [74].

Qing et al. [76] calculated the characteristics of gradient histogram (HOG), gabor and local binary pattern (LBP) and then Adaboost (adaptive boost) and SVM were used to define white-backed planthoppers in paddy fields and to define their developmental phases. Similarly, the writers used histograms of oriented gradient characteristics or region index as well as color indexes to design SVM structure in [77,78]; then SVM method is used to classify pests. In [79], discrete cosine transform is used to extract characteristics and then classify pests with ANN. The authors used the color function [81,82] to train a Gaussian mixture model (GMM), and for subsequent recognition, the trained model pest/diseases could be correctly segmented from the background.

There have been major developments in machine learning and deep convolutional neural networks (DCNN) [96,97]. Object recognition has been dramatically developed to enhance the precision of recognition of pests/diseases in farming. For example, He et al. used deep learning technology to train a DCNN of 54,306 disease images and healthy plant leaves collected under controlled conditions and tested 99.35% precision in their validation dataset [98]. Training datasets are also pioneering jobs [99].

Regardless of whether it is a monitored or unsupervised approach, the construction and testing of a recognition model require a large amount of pest/disease pictures. Pest/disease image collection, however, is hard and labor-intensive. By preprocessing technology, such as concentrating on or using a stick trap [75], researchers have tried to decrease the difficulty and the labor force. Such constructed algorithms, however, are not appropriate for pest/disease recognition in the natural farm scene.

Therefore pest/disease image acquisition technology is essential for the recognition of pests/diseases. We observed that little study is currently available to tackle this problem. A preprogrammed autonomous pest control robot for sampling *B. tabaci* adults was designed [100], but the image background is not yet active in a natural farm scene.

8.5.1 Acquisition data source of pyralidae insect

An automatic pest and disease detection and identification system collects the picture information used in this research. The system was built in Hainan University's technology application and demonstration area in the province of Hainan, China. The fundamental system design can be split into five main components: the camera sensor (automatic focusing, resolution 1600 u1200 and camera model KS2A01AF) and screen unit, trap unit, energy distribution unit, smart detection and recognition unit, and hardware bearer unit.

8.6 Power and fuel efficient robotics

Off-road engine-based cars use large amounts of fossil fuels that emit large amounts of pollution into the atmosphere. Internal combustion motors emit carbon dioxide (CO_2), nitrogen oxide (NO_x), carbon monoxide (CO), particulate matter, and hydrocarbons, according to the US Environmental Protection Agency (EPA) [101]. CO_2 and NO_x are greenhouse gases contributing to global warming, while emissions of sulfur dioxide (SO_2) and NO_x add to acid rain. The use of internal combustion engines is a significant problem for the environment. In addition, these chemical compounds trigger health issues as well. For example, NO_x can cause respiratory diseases and intensify existing heart disease; CO can reduce the delivery of oxygen to the tissues and organs of the body, reducing the ability to work, mental abilities, and the learning ability of an individual. Hydrocarbons are volatile organic compounds that, among other impacts, can trigger headaches, dizziness, and loss of awareness. In addition, some of these substances are carcinogenic and boost the probability of leukemia, such as benzene. Particulate matter emitted from combustion motors (nitrates, sulfates, organic chemicals, metals, and particles of dust) may also influence the functions of the lung and heart causing severe health issues.

Many attempts to mitigate these adverse impacts have carried out the analysis of the use of energy and pollution from agricultural tractors. Several studies compared distinct techniques in the early 2000s and calculated the average absolute and specific emission values from agricultural tractors, concluding that the use of hydrocarbon fuels must be gradually substituted by cleaner fuels or electrical systems [102]. Other studies suggested using a fossil fuel model to simulate feasible scenarios for agricultural manufacturing to enhance future techniques [103]. Researchers have studied in recent years how increasing soil organic carbon content reduces the plowing force draft, leading to lower energy consumption and emissions [104].

Several studies have evaluated exhaust gas emissions from internal combustion engines, and many of these researchers have concentrated on agricultural machines. For instance, a tractor mathematical model was created [105] to evaluate fuel consumption and engine emissions for various engine control strategies and engine transmission features, and the exhaust emissions and fuel composition of a true tractor during plowing were evaluated and correlated with the tractor load factor [106]. These works found that fuel consumption and emissions rely on the circumstances of engine velocity and load.

Many studies have evaluated the effect of alternative sources of energy such as biofuels and have showed that biofuels can benefit the environment and society [107]. Much of this research, however, has suggested the use of batteries and has examined multiple accessible battery technologies for use in solar-assisted hybrid electrical tractors that can be used in light-duty agricultural operations [108]. These scientists also performed life cycle analyses of a solar-assisted plugin hybrid electrical tractor and compared the outcomes to a comparable power output internal combustion engine tractor, taking into account both financial expenses and environmental emissions; they determined that the life cycle costs of solar-assisted hybrid electrical tractors are smaller than those of internal combustion engine tractors.

Fuel cells are another significant option for batteries. For instance, several scientists suggested the use of environmentally benign fuel cells for power generation in field crop manufacturing and distribution and submitted analyses of engineering systems showing how these systems can reduce pollution [109]. Others determined the net response by comparing the theoretical peak efficiencies of a fuel cell and a Carnot cycle [110]. They found that both systems have similar peak efficiencies, but a fuel cell is more efficient in practice because internal combustion engines cannot function at their theoretical maximum effectiveness. Other scientists compared electric vehicles for batteries, electric vehicles for hydrogen fuel cells, and hybrid plugin cars for hydrogen fuel cells [111]. These scientists have found comparable life cycle costs for electric battery cars and hydrogen fuel cell plugin hybrid cars. These vehicles' life cycle expenses are greater than internal combustion engine expenses but could be reduced by 2030. The strategy described in this work stemmed from the observation that the internal combustion engine often provided more energy than required during precision farming activities with robotized cars, especially when the agricultural tool or instrument used a power takeoff machine as an energy source. The aim of this job was, therefore, to create, execute, and evaluate a hybrid energy system for robotic agricultural vehicles. The suggested energy system combines the use of batteries, a hydrogen fuel cell, and photovoltaic cells with the tractor's initial internal combustion engine to obtain a significant reduction in fossil fuel consumption, decreasing the emissions of pollutants.

8.7 Conclusion

In this chapter, we discussed the recent application and development of agriculture robots. The study shows that most of these autonomous systems are more flexible than standard schemes and can considerably decrease labor costs, and limit the number of daily working hours. In addition, autonomous systems can be used to replace the most trivial working routines, although some routines are almost impossible to automate due to the accuracy of the specific tasks required. The Internet of Things, image processing, and navigation play vital roles in the development of agriculture robots. However, the original investments and annual expenses for costly GPS systems are still comparatively large at this stage of growth, but it appears possible to develop economically feasible robotic systems for grass cutting, crop scouting, and independent weeding. Findings indicate that if appropriate control and safety regulatory systems can be imposed at reasonable expenses, there is a substantial potential for implementing these systems. Furthermore, a comparison between distinct robotics applications shows that labor costs, crop rotation, and farm structure can have a tremendous effect on the future utilization of these technologies.

References

[1] C. Blanes, M. Mellado, C. Ortiz, A. Valera, Technologies for robot grippers in pick and place operations for fresh fruits and vegetables, Span. J. Agric. Res. 9 (4) (2011) 1130–1141.

[2] G. Muscato, M. Prestifilippo, A fuzzy-pd for the position and attitude control of an underwater robot, in: 2005 IEEE Conference on Emerging Technologies and Factory Automation, vol. 2, IEEE, 2005, 5 pp.

[3] R. Sam, S. Nefti, Design and development of flexible robotic gripper for handling food products, in: 2008 10th International Conference on Control, Automation, Robotics and Vision, IEEE, 2008, pp. 1684–1689.

[4] D. Galvez-Lopez, J.D. Tardos, Bags of binary words for fast place recognition in image sequences, IEEE Trans. Robot. 28 (5) (2012) 1188–1197.

[5] J. Artieda, J.M. Sebastian, P. Campoy, J.F. Correa, I.F. Mondragoon, C. Martinez, et al., Visual 3-d slam from uavs, J. Intell. Robotic Syst. 55 (4–5) (2009) 299.

[6] J. Libby, G. Kantor, Accurate gps-free positioning of utility vehicles for specialty agriculture, in: 2010 Pittsburgh, Pennsylvania, June 20-June 23, 2010, American Society of Agricultural and Biological Engineers, 2010, p. 1.

[7] T. Bakker, K. van Asselt, J. Bontsema, J. Muller, G. van Straten, Autonomous navigation using a robot platform in a sugar beet field, Biosyst. Eng. 109 (4) (2011) 357–368.

[8] H.W. Griepentrog, N.A. Andersen, J.C. Andersen, M. Blanke, O. Heinemann, T. Madsen, et al., Safe and reliable: further development of a field robot, Precis. Agric. 9 (2009) 857–866.

[9] L.P. Reis, F. Almeida, L. Mota, N. Lau, Coordination in multi-robot systems: applications in robotic soccer, in: International Conference on Agents and Artificial Intelligence, Springer, 2012, pp. 3–21.

[10] C.R. Turner, A. Fuggetta, L. Lavazza, A.L. Wolf, A conceptual basis for feature engineering, J. Syst. Softw. 49 (1) (1999) 3–15.

[11] J. Li, L. Chen, W. Huang, Q. Wang, B. Zhang, X. Tian, et al., Multispectral detection of skin defects of bi-colored peaches based on visnir hyperspectral imaging, Postharvest Biol. Technol. 112 (2016) 121–133.

[12] B. Zhang, W. Huang, J. Li, C. Zhao, S. Fan, J. Wu, et al., Principles, developments and applications of computer vision for external quality inspection of fruits and vegetables: a review, Food Res. Int. 62 (2014) 326–343.

[13] D. Longo, G. Muscato, Design and simulation of two robotic systems for automatic artichoke harvesting, Robotics 2 (4) (2013) 217–230.

[14] P.J. Sammons, T. Furukawa, A. Bulgin, Autonomous pesticide spraying robot for use in a greenhouse, in: Australian Conference on Robotics and Automation, vol. 1, 2005.

[15] A. Bechar, C. Vigneault, Agricultural robots for field operations: concepts and components, Biosyst. Eng. 149 (2016) 94–111.

[16] I.A. Hameed, A. la Cour-Harbo, O.L. Osen, Side-to-side 3d coverage path planning approach for agricultural robots to minimize skip/overlap areas between swaths, Robot. Auton. Syst. 76 (2016) 36–45.

[17] I.A. Hameed, Intelligent coverage path planning for agricultural robots and autonomous machines on three-dimensional terrain, J. Intell. Robotic Syst. 74 (3–4) (2014) 965–983.

[18] I.A. Hameed, D. Bochtis, C. Sorensen, Driving angle and track sequence optimization for operational path planning using genetic algorithms, Appl. Eng. Agric. 27 (6) (2011) 1077–1086.

[19] M. Hopkins, Automating in the 21st century career and technical education, Greenh. Grow. (2000) 4–12.

[20] T. Pilarski, M. Happold, H. Pangels, M. Ollis, K. Fitzpatrick, A. Stentz, The demeter system for automated harvesting, Auton. Robot. 13 (1) (2002) 9–20.

[21] R. Shamshiri, W.I.W. Ismail, A review of greenhouse climate control and automation systems in tropical regions, J. Agric. Sci. Appl. 2 (3) (2013) 176–183.

[22] S.L. Young, D.K. Giles, Targeted and microdose chemical applications, Automation: The Future of Weed Control in Cropping Systems, Springer, 2014, pp. 139–147.

[23] M.H. Ko, B.-S. Ryuh, K.C. Kim, A. Suprem, N.P. Mahalik, Autonomous greenhouse mobile robot driving strategies from system integration perspective: review and application, IEEE/ASME Trans. Mechatron. 20 (4) (2014) 1705–1716.

[24] A. Ruckelshausen, P. Biber, M. Dorna, H. Gremmes, R. Klose, A. Linz, et al., Bonirob-an autonomous field robot platform for individual plant phenotyping, Precis. Agric. 9 (841) (2009) 1.

[25] O. Bawden, D. Ball, J. Kulk, T. Perez, R. Russell, A lightweight, modular robotic vehicle for the sustainable intensification of agriculture. In: C, Chen (Ed.) Proceedings of the 16th Australasian Conference on Robotics and Automation 2014. Australian Robotics and Automation Association Inc., Australia, 2014, 1–9.

[26] A. Ruckelshausen, R. Klose, A. Linz, J. Marquering, M. Thiel, S. Tolke, Autonome roboter zur unkrautbekampfung, J. Plant. Dis. Protect. (2006) 173–180.

[27] H. Koselka, B. Wallach, Agricultural robot system and method, US Patent 7,854,108, December 21, 2010.

[28] R.N. Jorgensen, C.G. Sorensen, J. Maagaard, I. Havn, K. Jensen, H.T. Sogaard, et al., Hortibot: a system design of a robotic tool carrier for high-tech plant nursing, Agricult. Eng. Int. IX, 2007.

[29] O. Green, T. Schmidt, R.P. Pietrzkowski, K. Jensen, M. Larsen, R.N. Jorgensen, Commercial autonomous agricultural platform: Kongskilde robotti, in: Second International Conference on Robotics and associated High-technologies and Equipment for Agriculture and Forestry, RHEA, 2014, pp. 351–356.

[30] K. Jensen, M. Larsen, S. Nielsen, L. Larsen, K. Olsen, R. Jorgensen, Towards an open software platform for field robots in precision agriculture, Robotics 3 (2) (2014) 207–234.

[31] R. Bogue, Sensors key to advances in precision agriculture, Sens. Rev. 37 (1) (2017) 1–6.

[32] D.J. Mulla, Twenty five years of remote sensing in precision agriculture: key advances and remaining knowledge gaps, Biosyst. Eng. 114 (4) (2013) 358–371.

[33] R. Shamshiri, I. Wan, et al., Design and simulation of control systems for a field survey mobile robot platform, Res. J. Appl. Sci. Eng. Technol. 6 (13) (2013) 2307–2315.

[34] J. Hemming, C. Bac, B. van Tuijl, R. Barth, J. Bontsema, E. Pekkeriet, et al., A robot for harvesting sweet-pepper in greenhouses. Proceedings International Conference of Agricultural Engineering, Zurich, 2014, 1–8.

[35] P. Li, S.-h Lee, H.-Y. Hsu, Review on fruit harvesting method for potential use of automatic fruit harvesting systems, Procedia Eng. 23 (2011) 351–366.

[36] W. Qureshi, A. Payne, K. Walsh, R. Linker, O. Cohen, M. Dailey, Machine vision for counting fruit on mango tree canopies, Precis. Agric. 18 (2) (2017) 224–244.

[37] C. Weltzien, H.-H. Harms, N. Diekhans, Kfz radarsensor zur objekterkennung im landwirtschaftlichen umfeld, LANDTECHNIK-Agric. Eng. 61 (5) (2006) 250–251.

[38] K. Tanigaki, T. Fujiura, A. Akase, J. Imagawa, Cherry-harvesting robot, Comput. Electron. Agric. 63 (1) (2008) 65–72.

[39] D. Bulanon, T. Burks, V. Alchanatis, Study on temporal variation in citrus canopy using thermal imaging for citrus fruit detection, Biosyst. Eng. 101 (2) (2008) 161–171.

[40] H. Okamoto, W.S. Lee, Green citrus detection using hyperspectral imaging, Comput. Electron. Agric. 66 (2) (2009) 201–208.

[41] F. Westling, J. Underwood, S. Orn, Light interception modelling using unstructured lidar data in avocado orchards, Comput. Electron. Agric. 153 (2018) 177–187.

[42] D. Bulanon, T. Burks, V. Alchanatis, Image fusion of visible and thermal images for fruit detection, Biosyst. Eng. 103 (1) (2009) 12–22.

[43] A. King, et al., The future of agriculture, Nature 544 (7651) (2017) S21–S23.

[44] S. Wolfert, L. Ge, C. Verdouw, M.-J. Bogaardt, Big data in smart farming—a review, Agric. Syst. 153 (2017) 69–80.

[45] A. Chlingaryan, S. Sukkarieh, B. Whelan, Machine learning approaches for crop yield prediction and nitrogen status estimation in precision agriculture: a review, Comput. Electron. Agric. 151 (2018) 61–69.

[46] U. Drach, I. Halachmi, T. Pnini, I. Izhaki, A. Degani, Automatic herding reduces labour and increases milking frequency in robotic milking, Biosyst. Eng. 155 (2017) 134–141.

[47] A. Bach, V. Cabrera, Robotic milking: feeding strategies and economic returns, J. Dairy. Sci. 100 (9) (2017) 7720–7728.

[48] M. Gonzalez-de Soto, L. Emmi, M. Perez-Ruiz, J. Aguera, P. Gonzalez-de Santos, Autonomous systems for precise spraying-evaluation of a robotised patch sprayer, Biosyst. Eng. 146 (2016) 165–182.

[49] G. Adamides, C. Katsanos, Y. Parmet, G. Christou, M. Xenos, T. Hadzilacos, et al., Hri usability evaluation of interaction modes for a teleoperated agricultural robotic sprayer, Appl. Ergon. 62 (2017) 237–246.

[50] L. Comba, P. Gay, D.R. Aimonino, Robot ensembles for grafting herbaceous crops, Biosyst. Eng. 146 (2016) 227–239.

[51] C. Zhang, H. Gao, J. Zhou, A. Cousins, M.O. Pumphrey, S. Sankaran, 3d robotic system development for high-throughput crop phenotyping, IFAC-PapersOnLine 49 (16) (2016) 242–247.

[52] R. Barth, J. Hemming, E.J. van Henten, Design of an eye-in-hand sensing and servo control framework for harvesting robotics in dense vegetation, Biosyst. Eng. 146 (2016) 71–84.

[53] J. De Baerdemaeker, A. Munack, H. Ramon, H. Speckmann, Mechatronic systems, communication, and control in precision agriculture, IEEE Control Syst. Mag. 21 (5) (2001) 48–70.

[54] R.R. Shamshiri, F. Kalantari, K. Ting, K.R. Thorp, I.A. Hameed, C. Weltzien, et al., Advances in greenhouse automation and controlled environment agriculture: a transition to plant factories and urban agriculture, Int. J. Agricult. Biol. Eng. 11 (1) (2018) 1–22.

[55] J. Vazquez, E. Lacarra, M. Sanchez, J. Rioja, J. Bruzual, Edas (egnos data access service): differential gps corrections performance test with state-of-the-art precision agriculture system, in: 30th International Technical Meeting of The Satellite-Division-ofthe-Institute-of-Navigation (ION GNSS +), pp. 1988–1998.

[56] L. Yang, D. Gao, Y. Hoshino, S. Suzuki, Y. Cao, S. Yang, Evaluation of the accuracy of an auto-navigation system for a tractor in mountain areas, in: 2017 IEEE/SICE International Symposium on System Integration (SII), IEEE, 2017, pp. 133–138.

[57] T.J. Esau, Q.U. Zaman, Y.K. Chang, A.W. Schumann, D.C. Percival, A.A. Farooque, Spot-application of fungicide for wild blueberry using an automated prototype variable rate sprayer, Precis. Agric. 15 (2) (2014) 147–161.

[58] D. Ball, B. Upcroft, G. Wyeth, P. Corke, A. English, P. Ross, et al., Vision-based obstacle detection and navigation for an agricultural robot, J. Field Robot. 33 (8) (2016) 1107–1130.

[59] G. Jiang, Z. Wang, H. Liu, Automatic detection of crop rows based on multi-rois, Expert. Syst. Appl. 42 (5) (2015) 2429–2441.

[60] M. Kragh, R.N. Jorgensen, H. Pedersen, Object detection and terrain classification in agricultural fields using 3d lidar data, in: International Conference on Computer Vision Systems, Springer, 2015, pp. 188–197.

[61] P. Bernad, P. Lepej, C. Rozman, K. Pazek, J. Rakun, An evaluation of three different infield navigation algorithms, in: Agricultural Robots-Fundamentals and Applications, IntechOpen, 2018.

[62] S. Kohlbrecher, J. Meyer, T. Graber, K. Petersen, U. Klingauf, O. von Stryk, Hector open source modules for autonomous mapping and navigation with rescue robots, Robot Soccer World Cup, Springer, 2013, pp. 624–631.

[63] P. Lepej, J. Rakun, Simultaneous localisation and mapping in a complex field environment, Biosyst. Eng. 150 (2016) 160–169.

[64] P. Lepej, J. Maurer, S. Uran, G. Steinbauer, Dynamic arc fitting path follower for skid-steered mobile robots, Int. J. Adv. Robotic Syst. 12 (10) (2015) 139.

[65] T. Shen, S. Radmard, A. Chan, E.A. Croft, G. Chesi, Optimized vision-based robot motion planning from multiple demonstrations, Auton. Robot. 42 (6) (2018) 1117–1132.

[66] M. Quigley, B. Gerkey, W.D. Smart, Programming Robots With ROS: A Practical Introduction to the Robot Operating System, O'Reilly Media, Inc, 2015.

[67] L. Zhao, J. Ding, Least squares approximations to lognormal sum distributions, IEEE Trans. Vehic. Technol. 56 (2) (2007) 991–997.

[68] P. Lepej, Sistemi in krmiljenje avtonomnih mobilnih robotov v neurejenih prostorih (Ph.D. thesis), Univerza v Mariboru (Slovenia), 2018.

[69] T. Wei, W. Zhu, M. Pang, Y. Liu, Z. Wang, J.-G. Dong, Influence of the damage of cotton bollworm and corn borer to ear rot in corn, J. Maize Sci. 21 (4) (2013) 116–118.

[70] C. Wen, D. Guyer, Image-based orchard insect automated identification and classification method, Comp. Electron. Agric. 89 (2012) 110–115.

[71] S.R. Rupanagudi, B. Ranjani, P. Nagaraj, V.G. Bhat, G. Thippeswamy, A novel cloud computing based smart farming system for early detection of borer insects in tomatoes, in: 2015 International Conference on Communication, Information & Computing Technology (ICCICT), IEEE, 2015, pp. 1–6.

[72] H. Ali, M. Lali, M.Z. Nawaz, M. Sharif, B. Saleem, Symptom based automated detection of citrus diseases using color histogram and textural descriptors, Comput. Electron. Agric. 138 (2017) 92–104.

[73] J. Lu, R. Ehsani, Y. Shi, J. Abdulridha, A.I. de Castro, Y. Xu, Field detection of anthracnose crown rot in strawberry using spectroscopy technology, Comput. Electron. Agric. 135 (2017) 289–299.

[74] C. Xie, C. Yang, Y. He, Hyperspectral imaging for classification of healthy and gray mold diseased tomato leaves with different infection severities, Comput. Electron. Agric. 135 (2017) 154–162.

[75] C. Xia, J.-M. Lee, Y. Li, B.-K. Chung, T.-S. Chon, In situ detection of small-size insect pests sampled on traps using multifractal analysis, Opt. Eng. 51 (2) (2012) 027001.

[76] Y. Qing, G.-T. Chen, W. Zheng, C. Zhang, B.-J. Yang, T. Jian, Automated detection and identification of white-backed planthoppers in paddy fields using image processing, J. Integr. Agric. 16 (7) (2017) 1547–1557.

[77] T. Liu, W. Chen, W. Wu, C. Sun, W. Guo, X. Zhu, Detection of aphids in wheat fields using a computer vision technique, Biosyst. Eng. 141 (2016) 82–93.

[78] M. Ebrahimi, M. Khoshtaghaza, S. Minaei, B. Jamshidi, Vision-based pest detection based on svm classification method, Comput. Electron. Agric. 137 (2017) 52–58.

[79] M.H. Javed, M.H. Noor, B.Y. Khan, N. Noor, T. Arshad, K-means based automatic pests detection and classification for pesticides spraying, Int. J. Adv. Comput. Sci. Appl. 8 (11) (2017) 236–240.

[80] S. Dai, H. Man, A convolutional riemannian texture model with differential entropic active contours for unsupervised pest detection, in: 2017 IEEE International Conference on Acoustics, Speech and Signal Processing (ICASSP), IEEE, 2017, pp. 1028–1032.

[81] Y. Zhao, Z. Hu, B. Yong, et al., An accurate segmentation approach for disease and pest based on texture difference guided drlse, Trans. Chin. Soc. Agric. Mach. 46 (2) (2015) 14–19.

[82] Y. Zhao, Z. Hu, Segmentation of fruit with diseases in natural scenes based on logarithmic similarity constraint otsu, Trans. Chin. Soc. Agric. Mach. 46 (11) (2015) 9–15.

[83] Y.-Q. Hu, L.-T. Song, J. Zhang, C.-J. Xie, R. Li, Pest image recognition of multi-feature fusion based on sparse representation, Pattern Recognit. Artif. Intell. 27 (11) (2014) 985–992.

[84] M. Bayat, M. Abbasi, A. Yosefi, Improvement of pest detection using histogram adjustment method and gabor wavelet, J. Asian Sci. Res. 6 (2) (2016) 24.

[85] A. Martin, D. Sathish, C. Balachander, T. Hariprasath, G. Krishnamoorthi, Identification and counting of pests using extended region grow algorithm, in: 2015 2nd International Conference on Electronics and Communication Systems (ICECS), IEEE, 2015, pp. 1229–1234.

[86] L. Przybylowicz, M. Pniak, A. Tofilski, Semiautomated identification of european corn borer (lepidoptera: Crambidae), J. Econ. Entomol. 109 (1) (2015) 195–199.

[87] L. Deng, R. Yu, Pest recognition system based on bio-inspired filtering and lcp features, in: 2015 12th International Computer Conference on Wavelet Active Media Technology and Information Processing (ICCWAMTIP), IEEE, 2015, pp. 202–204.

[88] J.L. Miranda, B.D. Gerardo, B.T. Tanguilig, Pest identification using image processing techniques in detecting image pattern through neural network, in: Conference on Advances in Computer and Electronics Technology-ACET, 2014.

[89] P. Boniecki, K. Koszela, H. Piekarska-Boniecka, J. Weres, M. Zaborowicz, S. Kujawa, et al., Neural identification of selected apple pests, Comput. Electron. Agric. 110 (2015) 9–16.

[90] K. Espinoza, D.L. Valera, J.A. Torres, A. Lopez, F.D. Molina-Aiz, Combination of image processing and artificial neural networks as a novel approach for the identification of bemisia tabaci and frankliniella occidentalis on sticky traps in greenhouse agriculture, Comput. Electron. Agric. 127 (2016) 495–505.

[91] L. Zhu, Z. Zhang, P. Zhang, et al., Image identification of insects based on color histogram and dual tree complex wavelet transform (dtcwt), Acta Entomol. Sin. 53 (1) (2010) 91–97.

[92] L. Zhu, Z. Zhang, et al., Automatic insect classification based on local mean colour feature and supported vector machines, Orient. Insects 46 (3/4) (2012) 260–269.

[93] Z. Li, T. Hong, X. Zeng, J. Zheng, Citrus red mite image target identification based on k-means clustering, Trans. Chin. Soc. Agric. Eng. 28 (23) (2012) 147–153.

[94] A. Johannes, A. Picon, A. Alvarez-Gila, J. Echazarra, S. Rodriguez-Vaamonde, A.D. Navajas, et al., Automatic plant disease diagnosis using mobile capture devices, applied on a wheat use case, Comput. Electron. Agric. 138 (2017) 200–209.

[95] C. Xia, T.-S. Chon, Z. Ren, J.-M. Lee, Automatic identification and counting of small size pests in greenhouse conditions with low computational cost, Ecol. Inform. 29 (2015) 139–146.

[96] A. Krizhevsky, I. Sutskever, G.E. Hinton, Imagenet classification with deep convolutional neural networks, In: F. Pereira, C.J.C. Burges, L. Bottou and K.Q. Weinberger (Eds). Advances in Neural Information Processing Systems 25. Curran Associates, Inc. 2012, 1097–1105.

[97] K. Simonyan, A. Zisserman, Very deep convolutional networks for large-scale image recognition, arXiv preprint arXiv:1409.1556.

[98] K. He, X. Zhang, S. Ren, J. Sun, Deep residual learning for image recognition, in: Proceedings of the IEEE Conference on Computer Vision and Pattern Recognition, 2016, pp. 770–778.

[99] S. Yang, X. Yang, J. Mo, The application of unmanned aircraft systems to plant protection in china, Precis. Agric. 19 (2) (2018) 278–292.

[100] B.-K. Chung, C. Xia, Y.-H. Song, J.-M. Lee, Y. Li, H. Kim, et al., Sampling of bemisia tabaci adults using a pre-programmed autonomous pest control robot, J. Asia-Pacific Entomol. 17 (4) (2014) 737743.

[101] J.C. Kurtz, L.E. Jackson, W.S. Fisher, Strategies for evaluating indicators based on guidelines from the environmental protection agency's office of research and development, Ecol. Indic. 1 (1) (2001) 49–60.

[102] P.-A. Hansson, M. Lindgren, O. Noren, Pm—power and machinery: a comparison between different methods of calculating average engine emissions for agricultural tractors, J. Agric. Eng. Res. 80 (1) (2001) 37–43.

[103] T. Dalgaard, N. Halberg, J.R. Porter, A model for fossil energy use in danish agriculture used to compare organic and conventional farming, Agric. Ecosyst. Environ. 87 (1) (2001) 51–65.

[104] C. Peltre, T. Nyord, S. Bruun, L.S. Jensen, J. Magid, Repeated soil application of organic waste amendments reduces draught force and fuel consumption for soil tillage, Agric. Ecosyst. Environ. 211 (2015) 94–101.

[105] M. Lindgren, K. Arrhenius, G. Larsson, L. Bufver, H. Arvidsson, C. Wetterberg, et al., Analysis of unregulated emissions from an off-road diesel engine during realistic work operations, Atmos. Environ. 45 (30) (2011) 5394–5398.

[106] A. Janulevicius, A. Juostas, G. Pupinis, Tractor's engine performance and emission characteristics in the process of ploughing, Energy Convers. Manag. 75 (2013) 498–508.

[107] A. Gasparatos, P. Stromberg, K. Takeuchi, Biofuels, ecosystem services and human wellbeing: putting biofuels in the ecosystem services narrative, Agric. Ecosyst. Environ. 142 (3–4) (2011) 111–128.

[108] H. Mousazadeh, A. Keyhani, A. Javadi, H. Mobli, K. Abrinia, A. Sharifi, Life-cycle assessment of a solar assist plug-in hybrid electric tractor (sapht) in comparison with a conventional tractor, Energy Convers. Manag. 52 (3) (2011) 1700–1710.

[109] J.A. Mulloney Jr, Mitigation of carbon dioxide releases from power production via "sustainable agripower": the synergistic combination of controlled environmental agriculture (large commercial greenhouses) and disbursed fuel cell power plants, Energy Convers. Manag. 34 (911) (1993) 913–920.

[110] A.E. Lutz, R.S. Larson, J.O. Keller, Thermodynamic comparison of fuel cells to the carnot cycle, Int. J. Hydrog. Energy 27 (10) (2002) 1103–1111.

[111] G. Offer, D. Howey, M. Contestabile, R. Clague, N. Brandon, Comparative analysis of battery electric, hydrogen fuel cell and hybrid vehicles in a future sustainable road transport system, Energy Policy 38 (1) (2010) 24–29.

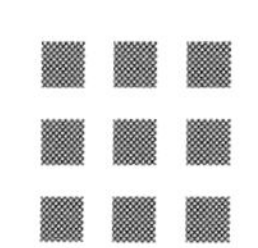

Index

Note: Page numbers followed by "*f*," "*t*," and "*b*" to refer to figures, tables, and boxes, respectively.

K

L

S

Printed in the United States
By Bookmasters